Computer Science, Technology and Applications

A Closer Look at Formation Control

COMPUTER SCIENCE, TECHNOLOGY AND APPLICATIONS

Additional books and e-books in this series can be found on Nova's website under the Series tab.

COMPUTER SCIENCE, TECHNOLOGY AND APPLICATIONS

A CLOSER LOOK AT FORMATION CONTROL

DIANWEI QIAN
EDITOR

nova
science publishers
New York

Library of Congress Cataloging-in-Publication Data

Names: Qian, Dianwei, editor.
Title: A closer look at formation control / Dianwei Qian, editor.
Identifiers: LCCN 2020028217 (print) | LCCN 2020028218 (ebook) |
ISBN 9781536181777 (hardcover) | ISBN 9781536182651 (adobe pdf)
Subjects: LCSH: Formation control (Machine theory) | Robots--Control systems.
Classification: LCC TJ211.35 .C56 2020 (print) | LCC TJ211.35 (ebook) | DDC 629.8/92--dc23
LC record available at https://lccn.loc.gov/2020028217
LC ebook record available at https://lccn.loc.gov/2020028218

Published by Nova Science Publishers, Inc. † New York

CONTENTS

PREFACE

Formation control is one of the most challenging problems in cooperative multi-robots. It is defined as a coordination of a group of robots to get into and to maintain a formation with a certain shape. The formation control problem has drawn significant attention for many years, and now it is well understood and tends to be mature. This control problem is originated from biological inspires such as flocking and schooling. Its classification includes formation shape generation, formation reconfiguration and selection, formation tracking, and role assignment in formation. It also has potential applications in search and rescue missions, forest fire detection and surveillance, etc. It can be extended to many realworld systems, autonomous robots, such as underwater vehicles, unmanned aerial vehicles, mobile sensor networks, rectangular agents, nonholonomic mobile robots, to name but a few.

Apparently, the book cannot include all research topics. The editor and the authors wish that it could reveal some tendencies on this research field and benefit readers. In this book, different aspects of formation control are explored. Chapters includes some new tendencies and developments in research on several formation methods of multi-robot systems, that is, the 1st-order sliding mode control, the 2nd-order sliding mode control, the integral sliding mode control, the terminal sliding mode control, the sliding model control of multi-agents and the fuzzy-based formation control of multi-quadrotor systems.

Chapter 1 – Coordination has emerged as one of central research themes in the field of robotics. There are some cooperative tasks like formation, coverage, and target aggregation. As a branch, formation control means to control a group of robots to form up and move in given patterns. Inspired by biological phenomena, the control problem rooted in the real applications is attractive. Many applications, such as search and rescue, surveillance, collaborative exploration, and exploration of large and unmapped areas, need the robots to maintain some prescribed formations when moving. Unfortunately, the formation tasks suffer from uncertainties. This chapter focuses on the modeling of the leader-follower formation scheme for multi-robot systems with uncertainties. The validity of the proposed model is addressed. According to the model, further tasks, e.g., performance analysis and control design in the presence of uncertainties, will be taken into consideration.

Chapter 2 – This chapter presents a control scheme for the leader-following formation of multiple robots. The scheme combines sliding mode and nonlinear disturbance observer. The formation dynamics are subject to formation uncertainties because of uncertain robotic individual. Concerning the formation uncertainties, the leader-following formation dynamics are modeled. Assuming that the upper bounds of the formation uncertainties are unknown, a nonlinear disturbance observer is designed. By selecting some parameters, the estimate error of the observer is exponentially convergent. Using the observer estimate, a sliding surface is investigated. Thereafter, a 1st-order sliding mode controller is designed in order to form up the robots. In the sense of Lyapunov, a sufficient condition is proven such that the formation system is asymptotically stable. Compared to the results by the sole SMC method, some numerical simulation results are presented to demonstrate the effectiveness and feasibility of the control scheme for multiple robots in the presence of uncertainties.

Chapter 3 – This chapter presents a 2nd-order sliding mode control method for the formation problem of multi-robot systems. The multiple robots contain plenty of uncertainties and disturbances. Such a control method is with two adaptive gains that can contribute to the robustness and improve the response of the formation maneuvers despite these uncertainties and disturbances. Based on the leader-follower frame, this control method is investigated. The closed-loop formation stability is theoretically guaranteed in the sense of Lyapunov. From the aspect of practice, the control method is carried out by a multi-robot system to achieve some desired formation patterns. Some numerical results are demonstrated to verify the feasibility of the control method. Some comparisons are also illustrated to support the superiority and effectiveness of the designed sliding mode control method.

Chapter 4 – This chapter investigates a coordinative scheme of integral sliding mode control for a group of nonholonomic mobile robots with mismatched uncertainties. Such a scheme is based on leader-follower approach to formation maneuvers for the robots. Since the models are subject to uncertainties, the mismatched uncertainties are challenging because it cannot be suppressed by the invariance of the integral sliding mode control. To suppress the mismatched uncertainties when forming up the robots, an integral sliding mode controller is designed. The system stability is also analyzed in the light of Lyapunov's theory and a sufficient condition is proven such that the formation maneuvers for the robots with mismatched uncertainties is of locally asymptotical stability in the sliding mode stage. To validate the feasibility of the controller, simulation results illustrate it can not only suppress the mismatched uncertainties, but also form up and maintain the robots in any trajectory Some numerical results are also conducted by a test bed made of three nonholonomic mobile robots.

Chapter 5 – This chapter investigates the formation problem of multiple robots based on the leader-following mechanism. At first, the dynamics of such a leader-follower framework are modeled. The input-output equations are depicted by calculating the relative degree of a leader-following formation system. Furthermore, the derivative and integral terminal sliding mode controller is designed based on the relative degree. Since the formation system suffers from the uncertainties, the nonlinear disturbance observer is adopted to deal with the uncertainties. The stability of the closed-loop control system is proven in the sense of Lynapunov. Finally, some numerical simulations are displayed to verify the feasibility and effectiveness by the designed controller and observer.

Chapter 6 – This chapter addresses a second-order sliding mode control method for the formation problem of multi-robot systems. This sliding mode control is based on the super-twisting law. In many real-world applications, the robots suffer from a great diversity of uncertainties and disturbances that challenge the super-twisting sliding mode formation maneuvers very much. Especially, such a challenge has the adverse effects on the formation performance when the uncertainties and disturbances have an unknown boundary. This chapter focuses on this issue and utilizes the technique of extreme learning machine to meet this challenge. Within the leader-follower framework, this chapter investigates the integration of the super-twisting sliding mode control method and the extreme learning machine. The output weights of this extreme learning machine are adaptively adjusted so that this integrated formation design has the guaranteed closed-loop stability in the sense of Lyaponov. In the end, some simulations are implemented via a multi-robot platform, which illustrate the superiority and effectiveness of the integrated formation design in spite of uncertainties and disturbances.

Chapter 7 – This chapter investigates the formation-control problem of multiple agents. The formation control is based on leader-following approaches. The method of integral sliding mode control is adopted to achieve formation maneuvers of the agents on the concept of graph theory. Since the agents are subject to uncertainties, the uncertainties also challenge the formation-control design. Under a mild assumption that the uncertainties have an unknown bound, the technique of nonlinear disturbance observer is utilized to attract the issue. According to a given communication topology, formation stability conditions are investigated by the observer-based integral sliding mode formation control. In the sense of Lyapunov, not only the formation stability is guaranteed, but the desired formation of the agents is also realized. Finally, some simulation results are presented to show the feasibility and validity of the proposed control scheme through a multi-agent platform.

Chapter 8 – Multi-robot systems have played essential roles in both civilian and military domains. The task assigned to the robotic systems has become more complicated and time-consuming to be carried out by a single robot. To reduce the costs and to improve the robustness and the efficacy of the system, formation control approaches for networked robots are highly desirable e.g., for a team of Unmanned Aerial Vehicles (UAVs). However, current state-of-the-art algorithms are faced with several challenging issues due to the presence of uncertainties such as wind gusts. Since traditional PID controllers or other fixed-gain control systems may not lead to satisfactory outcomes, this chapter presents a novel adaptive Strictly Negative Imaginary (SNI) formation control strategy based on knowledge-based Fuzzy Inference Systems to deal with the challenging dynamics of flight environments. To facilitate the information sharing process, the decentralized control architecture is designed. Each UAV only needs to measure its relative position with respect to its neighbours with a sensor such as a laser range-finder or machine vision system. Consequently, the required formation is formed by maintaining the desired relative positions between UAVs. The chapter rigorously compares the performance of the adaptive SNI control systems with respect to the effectiveness of the conventional PID controllers. The research highlights the efficacy of the proposed adaptive control systems, not only using numerical simulations, but also by conducting real-time flight tests.

Dianwei Qian
School of Control and Comupter Engineering
North China Electric Power University
Beijing, China

In: A Closer Look at Formation Control
Editor: Dianwei Qian
ISBN: 978-1-53618-177-7

Chapter 1

MODELING OF LEADER-FOLLOWER FORMATIONS

Dianwei Qian *

School of Control and Computer Engineering,
North China Electric Power University, Beijing, China

ABSTRACT

Coordination has emerged as one of central research themes in the field of robotics. There are some cooperative tasks like formation, coverage, and target aggregation. As a branch, formation control means to control a group of robots to form up and move in given patterns. Inspired by biological phenomena, the control problem rooted in the real applications is attractive. Many applications, such as search and rescue, surveillance, collaborative exploration, and exploration of large and unmapped areas, need the robots to maintain some prescribed formations when moving. Unfortunately, the formation tasks suffer from uncertainties. This chapter focuses on the modeling of the leader-follower formation scheme for multi-robot systems with uncertainties. The validity of the proposed model is addressed. According to the model, further

* Corresponding Author's Email:dianwei.qian@ncepu.edu.cn.

tasks, e.g., performance analysis and control design in the presence of uncertainties, will be taken into consideration.

Keywords: formation, modeling, leader-follower, multiple robots

INTRODUCTION

Compared with a single complex robot, the ability to coordinate swarms of multiple simple robots is of merit, including but not limited to efficiency, effectiveness and economy [1, 2]. Coordination of multiple mobile robots has emerged as one of the central research themes [3]. In the recent years, the cooperation of multiple robots has been extensively researched, which arises from the broad potential applications including many military applications such as general marine mine sweeping, exploration, surveillance and tracking, and many civilian applications such as mining support, search and rescue, farm aerial spraying, timber survey, forest fire suppression, etc [4−7].

Inspired by biological phenomena such as flocks of birds in flight, schools of fishes in swimming, team of ants in huntting and swarms of bees in collecting, this control problem is attractive because it is rooted in the real world [8, 9]. Many applications, such as search and rescue, surveillance, collaborative exploration and exploration of large and unmapped areas, need the robots to maintain some prescribed formations when moving [10−13].

Several coordinated control schemes have been reported for multi-robot formations. These schemes can be roughly categorized into behavior-based techniques, virtual-structure-based methods, graph-based and leader-follower-based approaches. All the three categories have both benefits and drawbacks [14, 15].

In the behavior-based approach, several desired behaviors (e.g., collision avoidance, formation keeping, target seeking) are prescribed for each robot, and the final control is derived from a weighting of the relative importance of each behavior [16]. The formation control is obtained from a weighted summation of each behavioral output. This approach allows a convenient generation of control strategies in the face of multiple competing objectives, and allows for an explicit feedback through communication between neighbors [17]. However, since the group behavior cannot be explicitly defined, and formal mathematical analysis on group stability is generally difficult, exact formation control is hard to guarantee. With this approach, it might be difficult to describe the dynamics of the group and guarantee the stability of the whole system [18, 19].

By treating the entire formation as a single entity, the virtual structure-based approach is comparatively more amenable to mathematical analysis [20]. Desired trajectories are not assigned to each single robot but to the entire formation as a whole. When the structure moves, it traces out desired trajectories for each vehicle in the group to track [21]. The behavior of the robot formation is predictable and consequently the stability analysis can be easily performed. It is fairly easy to prescribe the coordinated behavior for the group. The formation can be maintained very well during the maneuvers [22]. The virtual structure can evolve as a whole in a given direction with some given orientation and maintain a rigid geometric relationship among multiple vehicles. However, a large inter-robot communication bandwidth is required. When the formation shape is time-varying or needs to be frequently reconfigured, this approach may not be the optimal choice [23].

A formation is defined as a directed graph where the vertices represent the individual robots and the labeled edges represent geometric constraints that are maintained by the robots [24]. There can be a number of formation graphs that result in the same geometric arrangement. These graphs all belong to the same equivalence class. Compare with leader-follower approach, the unconstrained nodes of a formation graph are known as formation leaders. In this approach, the team size is scalable, just by adding nodes and edges [25]. Besides of control theory, we can use theories and techniques to solve the problem such as topology, data mining, etc. It is also easy to analyze the stability and can easy to reach stability. However, there is no relation with the dynamic and kinematic properties of robot. A large inter-robot communication bandwidth is required, and it is better to use with centralized structure.

Among them, the category of leader-follower-based approaches has been paid more and more attentions. The idea of the category is suggested by its name. A robot is designed as leader and other robots are named as followers. The leader moves along a predefined trajectory, while the followers keep track of the leader and maintain the desired distance and the desired relative bearing angle [26]. It is apparent that the category is hierarchical. The main criticism is that such a hierarchical structure heavily depends on the leader to achieve formation maneuvers. Nevertheless, the category has become popular because a reference trajectory of the leader can be clearly defined and the formation stability of the followers can be exactly guaranteed [27].

With the leader-follower formation strategy, there is defined a group leader which leads the group bulk motion. The other robots are the followers that maintain the respective relationships with the group leader robot, in general. However, when the number of mobile robots in the group is large, the relationships of some follower robot with leader robot are hard to define due to the limitation of sensors' working range. In this book, only the information of nearby robots is available to each robot. Therefore, the definition of whole group relationships is a combination of relationships which include one follower and one leader based on desired distance-desired bearing, one follower and two leaders based on desired distance-desired distance. When one follower robot follows two leader robots, the leader robots are the robots nearby the follower robot for easy sensorial connection.

MODELING OF A SINGLE ROBOT

There exists a unicycle-like robot in Figure 1.1. The robot moves in the horizontal plane. It is round and the diameter is $2r$. Its two parallel wheels have a same axis, independently controlled by two direct current motors. The robot can simultaneously rotate and translate, described by

$$\mathbf{q} = \begin{bmatrix} x & y & \theta \end{bmatrix}^T \tag{1.1}$$

In (1), (*x*, *y*) is located at the center of the robot and represents its translational coordinates, and θ indicates its rotational coordinate. To know the position, a positioning sensor at the front castor of this robot is set up in Figure 1.1. The axis of the sensor is orthogonal to the axis of the two wheels.

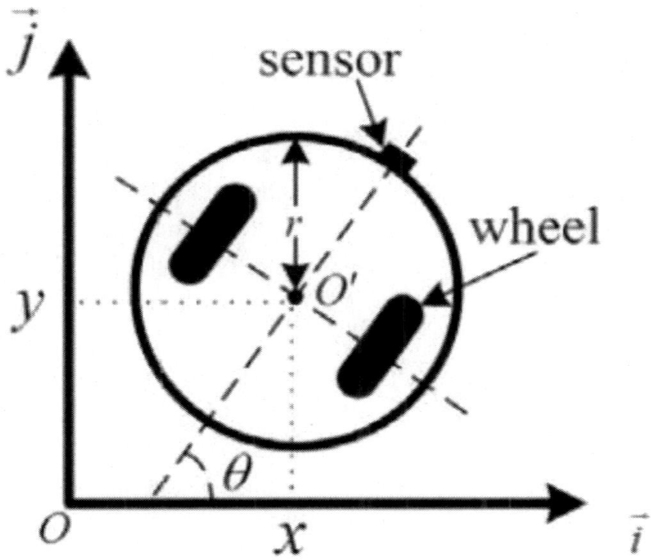

Figure 1.1. Sketches of the unicycle-like robot.

On the assumption of pure rolling and no-slipping, the ideal kinematic model of this robot [26, 27] has the form of

$$\dot{\mathbf{q}} = \begin{bmatrix} \dot{x} \\ \dot{y} \\ \dot{\theta} \end{bmatrix} = \begin{bmatrix} \cos\theta & 0 \\ \sin\theta & 0 \\ 0 & 1 \end{bmatrix} \cdot \begin{bmatrix} v \\ \omega \end{bmatrix} \tag{1.2}$$

$$\text{s.t. } \dot{x}\sin\theta - \dot{y}\cos\theta = 0 \tag{1.3}$$

Here v is the robot's linear velocity in the X-Y coordinates and its direction is determined by the X-Y coordinates as well. ω represents the angular velocity and its direction is positive when the robot rotates counterclockwise.

Concerning the constraint (1.3), the time derivative of (1.2), namely, the ideal dynamic model, can be written as

$$\begin{bmatrix} \ddot{x} \\ \ddot{y} \\ \ddot{\theta} \end{bmatrix} = \begin{bmatrix} -\dot{y}\dot{\theta} \\ \dot{x}\dot{\theta} \\ 0 \end{bmatrix} + \begin{bmatrix} \cos\theta & 0 \\ \sin\theta & 0 \\ 0 & 1 \end{bmatrix} \cdot \mathbf{u} \tag{1.4}$$

In (1.4), $\mathbf{u} = [\dot{v} \quad \dot{\omega}]^T$. Here the derivatives of v and ω represent the acceleration and angular acceleration of the robot, respectively. Since the robot in reality suffers from a variety of uncertainties and disturbances, for example, friction, slip and slide shift, and so forth, the real dynamic model can be derived from (1.4).

$$\begin{bmatrix} \ddot{x} \\ \ddot{y} \\ \ddot{\theta} \end{bmatrix} = \begin{bmatrix} -\dot{y}\dot{\theta} \\ \dot{x}\dot{\theta} \\ 0 \end{bmatrix} + \begin{bmatrix} \cos\theta & 0 \\ \sin\theta & 0 \\ 0 & 1 \end{bmatrix} \cdot (\mathbf{u} + \Delta \cdot \mathbf{u}) + \boldsymbol{\pi}(\mathbf{q}, \dot{\mathbf{q}}) \tag{1.5}$$

In (1.5), the term $\boldsymbol{\pi}(\mathbf{q}, \dot{\mathbf{q}})$ represents the lumped uncertainties and disturbances, defined by

$$\boldsymbol{\pi}(\mathbf{q}, \dot{\mathbf{q}}) = [\pi_x \quad \pi_y \quad \pi_\theta]^T$$

Here π_x π_y and π_θ are the functions of the vectors $\mathbf{q}$ and $\dot{\mathbf{q}}$. $\boldsymbol{\Delta}$ indicates the physical parameter changes of this robot, described by

$$\boldsymbol{\Delta} = \begin{bmatrix} \varepsilon & 0 \\ 0 & \varepsilon' \end{bmatrix}$$

here ε and ε' are the changes of the mass and the inertia of the robot, respectively.

MODELING OF A LEADER-FOLLOWER PAIR

Consider a multi-robot system containing N robots. Each robot is the same as the robot in Figure 1.1. Without loss of generality, the robot i is selected as the leader and it makes up N−1 leader-follower pairs with the left robots. Figure 1.2 illustrates such a leader-follower pair made of the leader i and the follower k.

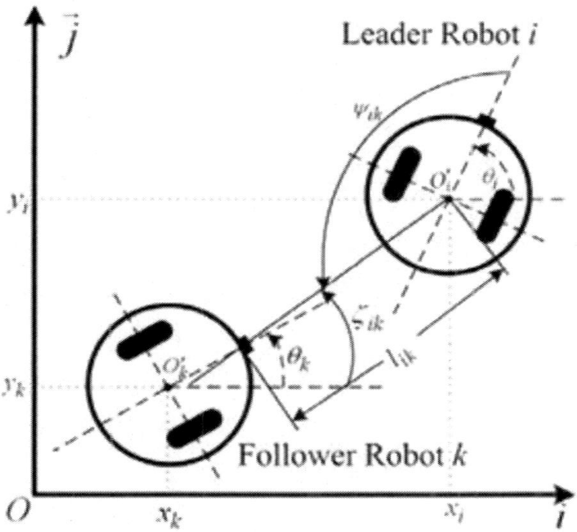

Figure 1.2. Sketches of a leader-follower pair.

In Figure 1.2, the subscripts i is adopted to label the individual variables of the leader, the subscripts k is employed to describe the individual variables of the follower and the subscript ik is used for the relative variables of this pair. Here the relative distance l_{ik} means the distance between the leader's center and the follower's front castor, formulated by

$$l_{ik} = \sqrt{\left(x_i - \bar{x}_k\right)^2 + \left(y_i - \bar{y}_k\right)^2} \tag{1.6}$$

Here

$$\bar{x}_k = x_k + r\cos\theta_k$$
$$\bar{y}_k = y_k + r\sin\theta_k$$

The relative bearing angle ψ_{ik} of the leader-follower pair is determined by

$$\psi_{ik} = \pi + \zeta_{ik} - \theta_i \tag{1.7}$$

Here

$$\zeta_{ik} = \arctan\frac{y_i - y_k - r\sin\theta_k}{x_i - x_k - r\cos\theta_k}$$

Differentiate l_{ik} and ψ_{ik} in (1.6) and (1.7}) twice with respect to time t, respectively. Dynamical equations of the pair based on the leader-follower scheme have the form of

$$\ddot{\mathbf{h}}_{ik} = \mathbf{G}(\mathbf{I}_2 + \Delta_k)\mathbf{u}_k + \mathbf{L}(\mathbf{I}_2 + \Delta_i)\mathbf{u}_i + \tilde{\mathbf{F}} + \tilde{\mathbf{P}} \tag{1.8}$$

where $\mathbf{h}_{ik} = \left[l_{ik}\ \ \psi_{ik}\right]^T$, $\mathbf{I}_2$ denotes a 2×2 identity matrix, $\mathbf{P}$ reflects the disturbances and the parameter variations. $\mathbf{G} = \begin{bmatrix} \cos\varphi_{ik} & r\sin\varphi_{ik} \\ -\dfrac{\sin\varphi_{ik}}{l_{ik}} & \dfrac{r\cos\varphi_{ik}}{l_{ik}} \end{bmatrix}$ here $\varphi_{ik} = \psi_{ik} + \theta_{ik}$ and $\theta_{ik} = \theta_i - \theta_k$. Further, $\mathbf{L}$ is described by $\mathbf{L} = \begin{bmatrix} -\cos\psi_{ik} & 0 \\ \dfrac{\sin\psi_{ik}}{l_{ik}} & -1 \end{bmatrix}$. $\tilde{\mathbf{F}} = \left[\tilde{F}_1\ \tilde{F}_2\right]^T$, where $\tilde{F}_1$ and $\tilde{F}_2$ are given by

$$\tilde{F}_1 = (\dot{\psi}_{ik})^2 l_{ik} + 2\dot{\psi}_{ik}\dot{\theta}_i l_{ik} + (\dot{\theta}_i)^2 l_{ik} - r\cos\phi_{ik}(\dot{\theta}_k)^2$$
$$-(\dot{y}_k\dot{\theta}_k - \dot{y}_i\dot{\theta}_i)\cos(\psi_{ik} + \theta_i) - (\dot{x}_i\dot{\theta}_i - \dot{x}_k\dot{\theta}_k)\sin(\psi_{ik} + \theta_i)$$

$$\tilde{F}_2 = -\frac{(\dot{y}_k\dot{\varphi}_{ik} - \dot{\psi}_{ik}\dot{y}_i)\sin(\psi_{ik} + \theta_i)}{l_{ik}} - \frac{(\dot{x}_k\dot{\varphi}_{ik} - \dot{\psi}_{ik}\dot{x}_i)\cos(\psi_{ik} + \theta_i) + r\dot{\theta}_k\dot{\varphi}_{ik}\sin\varphi_{ik}}{l_{ik}}$$
$$+\dot{l}_{ik}\frac{(\dot{y}_i - \dot{y}_k)\cos(\psi_{ik} + \theta_i)}{l_{ik}^2} - \dot{l}_{ik}\frac{(\dot{x}_i - \dot{x}_k)\sin(\psi_{ik} + \theta_i) + r\dot{\theta}_k\cos\varphi_{ik}}{l_{ik}^2}$$

$\mathbf{P} = \left[\tilde{P}_1 \;\; \tilde{P}_2\right]^T$ here $\tilde{P}_1$ and $\tilde{P}_2$ are written by

$$\tilde{P}_1 = -(\pi_{ix} - \pi_{kx})\cos(\psi_{ik} + \theta_i) - (\pi_{iy} - \pi_{ky})\sin(\psi_{ik} + \theta_i) + r\pi_{k\theta}\sin\varphi_{ik}$$

$$\tilde{P}_2 = \frac{(\pi_{ix} - \pi_{kx})\sin(\psi_{ik} + \theta_i) + r\pi_{k\theta}\cos\varphi_{ik}}{l_{ik}} - \frac{(\pi_{iy} - \pi_{ky})\cos(\psi_{ik} + \theta_i) + l_{ik}\pi_{i\theta}}{l_{ik}}$$

Conclusion

This chapter has modeled the leader-follower formation scheme for multi-robot systems with uncertainties. The validity of the proposed model is addressed. According to the proposed model, further tasks, e.g., performance analysis and control design in the presence of uncertainties, will be processed.

Appendix

According to $\zeta_{ik} = \arctan\dfrac{y_i - y_k - r\sin\theta_k}{x_i - x_k - r\cos\theta_k}$, (1.6) can be transformed as

$$x_i - x_k - r\cos\theta_k = l_{ik}\cos\zeta_{ik} \tag{1.9}$$

$$y_i - y_k - r\sin\theta_k = l_{ik}\sin\zeta_{ik} \tag{1.10}$$

Differentiating (1.9) and (1.10) with respect to time t yields

$$\dot{x}_i - \dot{x}_k + r\dot{\theta}_k \sin\theta_k = \dot{l}_{ik}\cos\zeta_{ik} - l_{ik}\dot{\zeta}_{ik}\sin\zeta_{ik} \tag{1.11}$$

$$\dot{y}_i - \dot{y}_k - r\dot{\theta}_k \cos\theta_k = \dot{l}_{ik}\sin\zeta_{ik} + l_{ik}\dot{\zeta}_{ik}\cos\zeta_{ik} \tag{1.12}$$

From (1.5), the dynamic equations of the leader i and the follower k can be depicted by

$$\begin{bmatrix}\ddot{x}_i\\ \ddot{y}_i\\ \ddot{\theta}_i\end{bmatrix} = \begin{bmatrix}-\dot{y}_i\dot{\theta}_i\\ \dot{x}_i\dot{\theta}_i\\ 0\end{bmatrix} + \begin{bmatrix}\cos\theta_i & 0\\ \sin\theta_i & 0\\ 0 & 1\end{bmatrix}\mathbf{u}_i + \begin{bmatrix}\cos\theta_i & 0\\ \sin\theta_i & 0\\ 0 & 1\end{bmatrix}\Delta_i\mathbf{u}_i + \begin{bmatrix}\pi_{ix}\\ \pi_{iy}\\ \pi_{i\theta}\end{bmatrix} \tag{1.13}$$

$$\begin{bmatrix}\ddot{x}_k\\ \ddot{y}_k\\ \ddot{\theta}_k\end{bmatrix} = \begin{bmatrix}-\dot{y}_k\dot{\theta}_k\\ \dot{x}_k\dot{\theta}_k\\ 0\end{bmatrix} + \begin{bmatrix}\cos\theta_k & 0\\ \sin\theta_k & 0\\ 0 & 1\end{bmatrix}\mathbf{u}_k + \begin{bmatrix}\cos\theta_k & 0\\ \sin\theta_k & 0\\ 0 & 1\end{bmatrix}\Delta_k\mathbf{u}_k + \begin{bmatrix}\pi_{kx}\\ \pi_{ky}\\ \pi_{k\theta}\end{bmatrix} \tag{1.14}$$

Differentiating (1.6), the first derivative of l_{ik} results in

$$\begin{aligned}\dot{l}_{ik} = \frac{1}{l_{ik}}[&(x_i - x_k - r\cos\theta_k)(\dot{x}_i - \dot{x}_k + r\dot{\theta}_k\sin\theta_k)\\ &+(y_i - y_k - r\sin\theta_k)(\dot{y}_i - \dot{y}_k - r\dot{\theta}_k\cos\theta_k)]\end{aligned} \tag{1.15}$$

Further, the second derivative of l_{ik} has the form of

$$\begin{aligned}\ddot{l}_{ik} = \frac{l_{ik}}{l_{ik}^2}\{&(\dot{x}_i - \dot{x}_k + r\dot{\theta}_k\sin\theta_k)^2 + (\dot{y}_i - \dot{y}_k - r\dot{\theta}_k\cos\theta_k)^2\\ &+(x_i - x_k - r\cos\theta_k)[\ddot{x}_i - \ddot{x}_k + r\ddot{\theta}_k\sin\theta_k + r\dot{\theta}_k^{\,2}\cos\theta_k]\\ &+(y_i - y_k - r\sin\theta_k)[\ddot{y}_i - \ddot{y}_k - r\ddot{\theta}_k\cos\theta_k + r\dot{\theta}_k^{\,2}\sin\theta_k]\}\\ &-\frac{\dot{l}_{ik}}{l_{ik}}\cdot\frac{\begin{array}{l}[(x_i - x_k - r\cos\theta_k)(\dot{x}_i - \dot{x}_k + r\dot{\theta}_k\sin\theta_k)\\ +(y_i - y_k - r\sin\theta_k)(\dot{y}_i - \dot{y}_k - r\dot{\theta}_k\cos\theta_k)]\end{array}}{l_{ik}}\end{aligned} \tag{1.16}$$

Substituting (1.9)--(1.14) into (1.16) yields

$$\begin{aligned}\ddot{l}_{ik} &= \cos\varphi_{ik}(1+\varepsilon_k)u_{k1} + r\sin\varphi_{ik}(1+\varepsilon_k')u_{k2} - \cos\psi_{ik}(1+\varepsilon_i)u_{i1} \\ &-[\cos\zeta_{ik}(\dot{y}_k\dot{\theta}_k - \dot{y}_i\dot{\theta}_i) + \sin\zeta_{ik}(\dot{x}_i\dot{\theta}_i - \dot{x}_k\dot{\theta}_k)] - r\dot{\theta}_k^2\cos\varphi_{ik} + l_{ik}\dot{\zeta}_{ik}^2 \\ &-[\cos\zeta_{ik}(\pi_{ix} - \pi_{kx}) + \sin\zeta_{ik}(\pi_{iy} - \pi_{ky})] + r\sin\varphi_{ik}\pi_{k\theta}\end{aligned} \tag{1.17}$$

Substituting $\zeta_{ik} = \psi_{ik} + \theta_i - \pi$ (transformed from (1.7)) into (1.17), the finial expression of $\ddot{l}_{ik}$ is determined by

$$\begin{aligned}\ddot{l}_{ik} &= \cos\varphi_{ik}(1+\varepsilon_k)u_{k1} + r\sin\varphi_{ik}(1+\varepsilon_k')u_{k2} - \cos\psi_{ik}(1+\varepsilon_i)u_{i1} \\ &+[-(\dot{y}_k\dot{\theta}_k - \dot{y}_i\dot{\theta}_i)\cos(\psi_{ik}+\theta_i) - (\dot{x}_i\dot{\theta}_i - \dot{x}_k\dot{\theta}_k)\sin(\psi_{ik}+\theta_i)] \\ &-r\dot{\theta}_k^2\cos\varphi_{ik} + l_{ik}(\dot{\psi}_{ik})^2 + 2l_{ik}\dot{\psi}_{ik}\dot{\theta}_i + l_{ik}(\dot{\theta}_i)^2 \\ &-(\pi_{ix} - \pi_{kx})\cos(\psi_{ik}+\theta_i) - (\pi_{iy} - \pi_{ky})\sin(\psi_{ik}+\theta_i) + r\sin\varphi_{ik}\pi_{k\theta}\end{aligned} \tag{1.18}$$

Differentiating (1.7), the first derivative of ψ_{ik} results in

$$\begin{aligned}\dot{\psi}_{ik} &= \dot{\zeta}_{ik} - \dot{\theta}_i \\ &= \frac{1}{1+\left(\frac{y_i - y_k - r\sin\theta_k}{x_i - x_k - r\cos\theta_k}\right)^2}\cdot\left(\frac{y_i - y_k - r\sin\theta_k}{x_i - x_k - r\cos\theta_k}\right)' - \dot{\theta}_i \\ &= \frac{(x_i - x_k - r\cos\theta_k)^2}{(x_i - x_k - r\cos\theta_k)^2 + (y_i - y_k - r\sin\theta_k)^2} \\ &\times\frac{[(\dot{y}_i - \dot{y}_k - r\dot{\theta}_k\cos\theta_k)(x_i - x_k - r\cos\theta_k) - (y_i - y_k - r\sin\theta_k)(\dot{x}_i - \dot{x}_k + r\dot{\theta}_k\sin\theta_k)]}{(x_i - x_k - r\cos\theta_k)^2} - \dot{\theta}_i \\ &= \frac{[(\dot{y}_i - \dot{y}_k - r\dot{\theta}_k\cos\theta_k)(x_i - x_k - r\cos\theta_k) - (y_i - y_k - r\sin\theta_k)(\dot{x}_i - \dot{x}_k + r\dot{\theta}_k\sin\theta_k)]}{l_{ik}^2} - \dot{\theta}_i\end{aligned} \tag{1.19}$$

Substituting (1.9) and (1.10) into (1.19) and taking $\theta_{ik} = \theta_i - \theta_k$ and $\varphi_{ik} = \psi_{ik} + \theta_{ik}$ into accounts, the expression of $\dot{\psi}_{ik}$ can be re-written as

$$\dot{\psi}_{ik} = \frac{(\dot{y}_i - \dot{y}_k)\cos\zeta_{ik} - (\dot{x}_i - \dot{x}_k)\sin\zeta_{ik} + r\dot{\theta}_k \cos\varphi_{ik}}{l_{ik}} - \dot{\theta}_i \tag{1.20}$$

Further, the second derivative of ψ_{ik} is determined by

$$\begin{aligned} \ddot{\psi}_{ik} = -\ddot{\theta}_i + \frac{\begin{matrix} [(\ddot{y}_i - \ddot{y}_k)\cos\zeta_{ik} - (\dot{y}_i - \dot{y}_k)\dot{\zeta}_{ik}\sin\zeta_{ik} \\ l_{ik}\ -(\ddot{x}_i - \ddot{x}_k)\sin\zeta_{ik} - (\dot{x}_i - \dot{x}_k)\dot{\zeta}_{ik}\cos\zeta_{ik} \\ +r\ddot{\theta}_k\cos\varphi_{ik} - r\dot{\theta}_k\dot{\varphi}_{ik}\sin\varphi_{ik}] \end{matrix}}{l_{ik}^2} \\ - \frac{\dot{l}_{ik}[(\dot{y}_i - \dot{y}_k)\cos\zeta_{ik} - (\dot{x}_i - \dot{x}_k)\sin\zeta_{ik} + r\dot{\theta}_k\cos\varphi_{ik}]}{l_{ik}^2} - r\dot{\theta}_i \end{aligned} \tag{1.21}$$

Substituting (1.13) and (1.14) into (1.21) results in

$$\begin{aligned} \ddot{\psi}_{ik} = & \frac{1}{l_{ik}} \begin{matrix} \{[\dot{x}_i\dot{\theta}_i + (1+\varepsilon_i)\sin\theta_i u_{i1} + \pi_{iy} - \dot{x}_k\dot{\theta}_k \\ -(1+\varepsilon_k)\sin\theta_k u_{k1} - \pi_{ky}]\cos\zeta_{ik} \\ -(\dot{y}_i - \dot{y}_k)\dot{\zeta}_{ik}\sin\zeta_{ik}\} \end{matrix} \\ & + \frac{1}{l_{ik}} \begin{matrix} \{[\dot{y}_i\dot{\theta}_i - (1+\varepsilon_i)\cos\theta_i u_{i1} - \pi_{ix} - \dot{y}_k\dot{\theta}_k + \\ (1+\varepsilon_k)\cos\theta_k u_{k1} + \pi_{kx}]\sin\zeta_{ik} \\ -(\dot{x}_i - \dot{x}_k)\dot{\zeta}_{ik}\cos\zeta_{ik}\} \end{matrix} \\ & + \frac{1}{l_{ik}} \begin{matrix} [r(1+\varepsilon_k')\cos\varphi_{ik}u_{k2} + d\cos\varphi_{ik}\pi_{k\theta} \\ -r\dot{\theta}_k\dot{\varphi}_{ik}\sin\varphi_{ik}] \end{matrix} \\ & - \frac{\dot{l}_{ik}}{l_{ik}^2} \begin{matrix} [(\dot{y}_i - \dot{y}_k)\cos\zeta_{ik} - (\dot{x}_i - \dot{x}_k)\sin\zeta_{ik} \\ +r\dot{\theta}_k\cos\phi_{ik}] \end{matrix} \\ & -(1+\varepsilon_i')u_{i2} - \pi_{i\theta} \end{aligned} \tag{1.22}$$

Further $\theta_i - \zeta_{ik} = \pi - \psi_{ik}$, $\zeta_{ik} = \psi_{ik} + \theta_i - \pi$ and $\theta_k - \zeta_{ik} = \pi - \phi_{ik}$ are satisfied. Substituting them into (1.22), the final expression of $\ddot{\psi}_{ik}$ can be written as

$$
\begin{aligned}
\ddot{\psi}_{ik} = & -\frac{(1+\varepsilon_k)\sin\varphi_{ik}}{l_{ik}}u_{k1} + \frac{r(1+\varepsilon_k')\cos\varphi_{ik}}{l_{ik}}u_{k2} \\
& + \frac{(1+\varepsilon_i)\sin\psi_{ik}}{l_{ik}}u_{i1} - (1+\varepsilon_i')u_{i2} \\
& - \frac{\begin{matrix}(\dot{\varphi}_{ik}\dot{y}_k - \dot{\psi}_{ik}\dot{y}_i)\sin(\psi_{ik}+\theta_i) \\ +(\dot{\varphi}_{ik}\dot{x}_k - \dot{\psi}_{ik}\dot{x}_i)\cos(\psi_{ik}+\theta_i) + r\dot{\theta}_k\dot{\varphi}_{ik}\sin\varphi_{ik}\end{matrix}}{l_{ik}} + \\
& \frac{\dot{l}_{ik}\begin{matrix}[(\dot{y}_i - \dot{y}_k)\cos(\psi_{ik}+\theta_i) \\ -(\dot{x}_i - \dot{x}_k)\sin(\psi_{ik}+\theta_i) - r\dot{\theta}_k\cos\varphi_{ik}]\end{matrix}}{l_{ik}^2} + \\
& \frac{\begin{matrix}(\pi_{ix} - \pi_{kx})\sin(\psi_{ik}+\theta_i) \\ -(\pi_{iy} - \pi_{ky})\cos(\psi_{ik}+\theta_i) + r\cos\varphi_{ik}\pi_{k\theta}\end{matrix}}{l_{ik}} - \pi_{i\theta}
\end{aligned} \tag{1.23}
$$

From (1.18) and (1.23), (1.8) is drawn to depict theleader-follower formation scheme for uncertain multi-robot systems.

REFERENCES

[1] Balch, T., Arkin, R. C. (1998). Behavior-based formation control for multirobot teams. *IEEE transactions on robotics and automation*, 14(6), 926-939.

[2] Oh, K. K., Park, M. C., Ahn, H. S. (2015). A survey of multi-agent formation control. *Automatica*, 53, 424-440.

[3] Beard, R. W., Lawton, J., Hadaegh, F. Y. (2001). A coordination architecture for spacecraft formation control. *IEEE Transactions on control systems technology*, 9(6), 777-790.

[4] Xiao, F., Wang, L., Chen, J., Gao, Y. (2009). Finite-time formation control for multi-agent systems. *Automatica*, 45(11), 2605-2611.

[5] Lewis, M. A., Tan, K. H. (1997). High precision formation control of mobile robots using virtual structures. *Autonomous robots*, 4(4), 387-403.

[6] Consolini, L., Morbidi, F., Prattichizzo, D., Tosques, M. (2008). Leader–follower formation control of nonholonomic mobile robots with input constraints. *Automatica*, 44(5), 1343-1349.
[7] Dong, X., Yu, B., Shi, Z., Zhong, Y. (2014). Time-varying formation control for unmanned aerial vehicles: Theories and applications. *IEEE Transactions on Control Systems Technology*, 23(1), 340-348.
[8] Defoort, M., Floquet, T., Kokosy, A., Perruquetti, W. (2008). Sliding-mode formation control for cooperative autonomous mobile robots. *IEEE Transactions on Industrial Electronics*, 55(11), 3944-3953.
[9] Shao, J., Xie, G., Wang, L. (2007). Leader-following formation control of multiple mobile vehicles. *IET Control Theory & Applications*, 1(2), 545-552.
[10] Do, K. D., Pan, J. (2007). Nonlinear formation control of unicycle-type mobile robots. *Robotics and Autonomous Systems*, 55(3), 191-204.
[11] Lin, Z., Wang, L., Han, Z., Fu, M. (2014). Distributed formation control of multi-agent systems using complex Laplacian. *IEEE Transactions on Automatic Control*, 59(7), 1765-1777.
[12] Gu, D. (2007). A differential game approach to formation control. *IEEE Transactions on Control Systems Technology*, 16(1), 85-93.
[13] Mariottini, G. L., Morbidi, F., Prattichizzo, D., Vander Valk, N., Michael, N., Pappas, G., Daniilidis, K. (2009). Vision-based localization for leader–follower formation control. *IEEE Transactions on Robotics*, 25(6), 1431-1438.
[14] Ge, X., Han, Q. L. (2017). Distributed formation control of networked multi-agent systems using a dynamic event-triggered communication mechanism. *IEEE Transactions on Industrial Electronics*, 64(10), 8118-8127.
[15] Dong, X., Zhou, Y., Ren, Z., Zhong, Y. (2016). Time-varying formation control for unmanned aerial vehicles with switching interaction topologies. *Control Engineering Practice*, 46, 26-36.

[16] Liao, F., Teo, R., Wang, J. L., Dong, X., Lin, F., Peng, K. (2016). Distributed formation and reconfiguration control of VTOL UAVs. *IEEE Transactions on Control Systems Technology*, 25(1), 270-277.

[17] Sun, Z., Mou, S., Anderson, B. D., Cao, M. (2016). Exponential stability for formation control systems with generalized controllers: A unified approach. *Systems & Control Letters*, 93, 50-57.

[18] Jin, X. (2016). Fault tolerant finite-time leader–follower formation control for autonomous surface vessels with LOS range and angle constraints. *Automatica*, 68, 228-236.

[19] Li, D., Ge, S. S., He, W., Ma, G., Xie, L. (2019). Multilayer formation control of multi-agent systems. *Automatica*, 109, 108558.

[20] De Queiroz, M., Cai, X., Feemster, M. (2019). *Formation Control of Multi-Agent Systems: A Graph Rigidity Approach*. John Wiley & Sons.

[21] Jin, L., Yu, S., Ren, D. (2019). Bearing-angle-based circular formation control of unicycles with arbitrary phase allocation. *The Journal of Engineering*, 2019(15), 623-628.

[22] Motoyama, T., Cai, K. (2019). Top-Down Synthesis of Multiagent Formation Control: An Eigenstructure Assignment Based Approach. *IEEE Transactions on Control of Network Systems*, 6(4), 1404-1414.

[23] Zhang, Q., Guo, H., Xiaohang, S. (2020). Formation Control of High-Order Swarm Systems With Time-Varying Delays and Switching Interconnections. *IEEE Access*, 8, 28188-28196.

[24] Qian, D., Zhang, G., Wang, J., Wu, Z. (2019). Second-Order Sliding Mode Formation Control of Multiple Robots by Extreme Learning Machine. *Symmetry*, 11(12), 1444.

[25] Kamalova, A., Navruzov, S., Qian, D., Lee, S. G. (2019). Multi-Robot exploration based on multi-objective grey wolf optimizer. *Applied Sciences*, 9(14), 2931.

[26] Qian, D., Tong, S., Li, C. (2016). Leader-Following Formation Control of Multiple Robots with Uncertainties through Sliding Mode and Nonlinear Disturbance Observer. *ETRI Journal*, 38(5), 1008-1018.

[27] Qian, D., Li, C., Lee, S., Ma, C. (2017). Robust formation maneuvers through sliding mode for multi-agent systems with uncertainties. *IEEE/CAA Journal of Automatica Sinica*, 5(1), 342-351.

In: A Closer Look at Formation Control
Editor: Dianwei Qian
ISBN: 978-1-53618-177-7

Chapter 2

1ST-ORDER SLIDING MODE-BASED FORMATION DESIGN

Dianwei Qian** and Fengyang Han
School of Control and Computer Engineering,
North China Electric Power University, Beijing, China

ABSTRACT

This chapter presents a control scheme for the leader-following formation of multiple robots. The scheme combines sliding mode and nonlinear disturbance observer. The formation dynamics are subject to formation uncertainties because of uncertain robotic individual. Concerning the formation uncertainties, the leader-following formation dynamics are modeled. Assuming that the upper bounds of the formation uncertainties are unknown, a nonlinear disturbance observer is designed. By selecting some parameters, the estimate error of the observer is exponentially convergent. Using the observer estimate, a sliding surface is investigated. Thereafter, a 1st-order sliding mode controller is designed in order to form up the robots. In the sense of Lyapunov, a sufficient condition is proven such that the formation system is asymptotically stable. Compared to the results by the sole SMC method, some numerical

* Corresponding Author's Email: dianwei.qian@ncepu.edu.cn.

simulation results are presented to demonstrate the effectiveness and feasibility of the control scheme for multiple robots in the presence of uncertainties.

Keywords: 1st-order sliding mode control, formation control, multiple robots, uncertainties

INTRODUCTION

Today, robots are changing the way we live and work [1]. In robotics, the formation-control problem of multiple robots is one of the most important fields [2]. To achieve formation maneuvers of multiple robots, various formation schemes have been presented [3, 4]. Among these schemes, the leader-following approaches have been paid more and more attentions [5]. Although they are criticized for 'single point of failure,' the leader-following approaches still become popular because they are simple enough to permit complete analyses and experiments, but the internal formation stability can be theoretically guaranteed [6−8].

Concerning leader-following approaches, the basic idea is that a multi-robot system is decomposed into several leader-follower pairs, where all followers share a leader. In each pair, leader takes charge of tracking a predefined trajectory while follower keeps track of the leader and maintains the desired distance and the desired relative angle [9, 10]. Adopting the idea, various control methods have been investigated for multi-robot systems, that is, robust control [11], predictive control [12], adaptive control [13], decentralized control feedback linearization [14] and sliding mode control [15], to name but a few.

As a nonlinear feedback design tool, the sliding mode control (SMC) method is robust against uncertainties [16, 17]. Especially, it is invariant to matched uncertainties [18]. There has been an increasing research interest in the SMC-based formation control. Some SMC-based techniques have been investigated for the formation-control problem, i.e., fast terminal SMC [19], adaptive fuzzy SMC [20], input-output linearization SMC [21], first-order SMC and second-order SMC [22]. In real applications, control systems inevitably suffer from uncertainties. Concerning the formation-control problem, some factors such as internal parameter fluctuations, unmodeled dynamics and external disturbances, result in formation uncertainties. The formation uncertainties not only deteriorate the formation performance, but also can they have adverse effects on the formation stability. Unfortunately, the formation uncertainties are inherently unmatched and cannot be suppressed by the invariance of SMC. Their existence greatly challenges the formation-control problem.

In previous works [19–22], the formation uncertainties are taken into account as well. To guarantee the formation stability, the formation uncertainties are assumed to be bounded by a known boundary. In the presence of uncertainties, such an assumption is not mild because the formation uncertainties are random and their boundary is hard to exactly know in advance. The lack of knowing the important information may cause several severe problems, i.e., decrease of the formation robustness, deterioration of the formation performance as far as deficiency of the formation stability.

The technique of nonlinear disturbance observer (NDOB) has been proven to be effective in handling uncertainties and improving robustness [23]. The applications of NDOB have been investigated by some actual cases [24–27]. Such a technique can be considered as an alternative to attack the issue of formation uncertainties [28]. So far, the academic problem of how to eliminate the adverse effects of the formation uncertainties by NDOB still remains problematic.

This chapter investigates a control scheme for the leader-following formation control of uncertain multiple robots. The scheme fuses SMC and NDOB together into a formation-control system. Under a mild assumption that formation uncertainties are just bounded, a NDOB is designed. By selecting some parameters, the estimate error of the observer is of exponential convergence. The observer output is adopted to construct a sliding surface. Thereafter, a sliding mode controller is developed by the sliding surface. A sufficient condition is derived from the Lyapunov direct method in order that the formation-control system has guaranteed stability in spite of uncertainties. To verify the feasibility and effectiveness of the control scheme, some simulation results are illustrated by multi-robot systems.

The remainder of this chapter is organized as follows. In Section 2, the leader-following formation dynamics are formulated in the presence of uncertainties. Combining SMC and NDOB, the control scheme in Section 3 is presented. In Section 4, the system stability is analyzed in the sense of Lyapunov. Finally, conclusions are drawn in Section 5.

MODELING THE LEADER-FOLLOWING FORMATION

The leader-following formation includes l–ψ and l–l formation frameworks. This chapter adopts the former framework. Take the multi-robot system into consideration. Without loss of generality, one robot is named leader and index i is utilized to describe the robot. The other N–1 robots are followers. These followers share the leader i such that the multi-robot system contains N–1 leader-follower pairs. Pick up one follower and index k is utilized to describe the follower. Then, the leader i and the follower k consists of a pair and the pair is employed to demonstrate the l–ψ formation framework.

Concerning such a leader-follower pair, some ideal conditions exist (1) there is no collision between the robots i and k; (2) there is no communication delay in the pair and (3) the follower k can measure its own posture as well as the posture of the leader i. The leader's duty is to track a predefined trajectory such that the leader's control problem can be treated as the tracking control of a single robot. As far as the follower k is concerned, its duty is to keep tracks of the leader i in accordance with the desired distance and the desired relative angle.

Recall Chapter 1, the mathematical model (1.8) can be described by

$$\ddot{\mathbf{h}}_{ik} = \mathbf{G}(\mathbf{I}_2 + \Delta_k)\mathbf{u}_k + \mathbf{L}(\mathbf{I}_2 + \Delta_i)\mathbf{u}_i + \tilde{\mathbf{F}} + \tilde{\mathbf{P}} \tag{2.1}$$

Here all the symbols are kept unchanged from Chapter 1.

In (2.1), we lump all uncertain terms in (2.1) and define the formation uncertainties as

$$\mathbf{d}_{ik} = \mathbf{G}\Delta_k \mathbf{u}_k + \mathbf{L}(\mathbf{I}_2 + \Delta_i)\mathbf{u}_i + \tilde{\mathbf{F}} + \tilde{\mathbf{P}} \tag{2.2}$$

Then, (2.1) can be re-arranged by

$$\ddot{\mathbf{h}}_{ik} = \mathbf{G}\mathbf{u}_k + \mathbf{d}_{ik} \tag{2.3}$$

Our motivation in the paper is to investigate a control scheme so that the follower k can trace the leader i despite the formation uncertainties.

From the aspect of the modern control theory, we define $\mathrm{x}_{ik} = [x_1\ x_2\ x_3\ x_4]^T$, where $x_1 = l_{ik}$, $x_2 = \dot{l}_{ik}$, $x_3 = \psi_{ik}$ and $x_4 = \dot{\psi}_{ik}$. Then, (2.3) can be re-written by

$$\dot{\mathbf{x}}_{ik} = \mathbf{A}_{ik}\mathbf{x}_{ik} + \mathbf{B}_{ik,1}\mathbf{u}_k + \mathbf{B}_{ik,2}\mathbf{d}_{ik} \tag{2.4}$$

where constant matrixes A_{ik} and $\mathrm{B}_{ik,2}$ are determined by $\begin{bmatrix} 0 & 1 & 0 & 0 \\ 0 & 0 & 0 & 0 \\ 0 & 0 & 0 & 1 \\ 0 & 0 & 0 & 0 \end{bmatrix}$ and $\begin{bmatrix} 0 & 0 \\ 1 & 0 \\ 0 & 0 \\ 0 & 1 \end{bmatrix}$, respectively; $\mathrm{B}_{ik,1}$ is a time-varying matrix, written by $\begin{bmatrix} 0 & 0 \\ \cos\varphi_{ik} & r\sin\varphi_{ik} \\ 0 & 0 \\ \frac{-\sin\varphi_{ik}}{l_{ik}} & \frac{r\cos\varphi_{ik}}{l_{ik}} \end{bmatrix}$.

FORMATION DESIGN

The SMC method is invariant to unmatched uncertainties. Unfortunately, the formation uncertainties are inherently unmatched. Described by d_{ik}, in (2.2), the lumped uncertain term consists of 4 parts, where $\mathrm{G}_{ik}\mathrm{D}_k\mathrm{u}_k$ is the sole matched part. Since the formation uncertainties contain random factors, their boundary is hard to exactly know in advance. Take the following mild assumption into consideration.

Assumption 1: d_{ik} is bounded by $\|d_{ik}\|_\infty \le d_{ik}^*$, here $d_{ik}^* > 0$ is unknown and $\|\bullet\|_\infty$ denotes ∞-norm.

Formation Design

Consider the formation dynamics (2.1). To develop a sliding mode controller to coordinate the leader-follower pair, the sliding surface vector (2.5) is defined by

$$\mathbf{s}_{ik} = \tilde{\mathbf{J}}_{ik}\mathbf{e}_{ik} + \tilde{\mathbf{K}}_{ik}\dot{\mathbf{e}}_{ik} \tag{2.5}$$

In (2.5), $s_{ik} = [s_{ik,1}\ s_{ik,2}]^T \in \Re^{2\times 1}$ is the sliding surface vector, $e_{ik} = [l_{ik} - l_{ik}^d\ \ \psi_{ik} - \psi_{ik}^d]^T \in \Re^{2\times 1}$ is the tracking error vector, l_{ik}^d and ψ_{ik}^d are the desired distance and the desired relative angle and $\tilde{\mathbf{J}}_{ik}$ and $\tilde{\mathbf{K}}_{ik}$ are 2×2 positive-definite matrixes to be designed, written by

$$\tilde{\mathbf{J}}_{ik} = \begin{bmatrix} \tilde{J}_1 & 0 \\ 0 & \tilde{J}_2 \end{bmatrix} \text{ and } \tilde{\mathbf{K}}_{ik} = \begin{bmatrix} \tilde{K}_1 & 0 \\ 0 & \tilde{K}_2 \end{bmatrix} \tag{2.6}$$

The 1st-order SMC law can be designed by

$$\mathbf{u}_k = -(\tilde{\mathbf{K}}_{ik}\mathbf{G})^{-1}[\tilde{\kappa}_{ik}\text{sign}(\mathbf{s}_{ik}) + \tilde{\eta}_{ik}\mathbf{s}_{ik} + \tilde{\mathbf{J}}_{ik}\mathbf{e}_{ik}] \tag{2.7}$$

here $\tilde{\kappa}_{ik}$ and $\tilde{\eta}_{ik}$ are positive constants and $\text{sign}(s_{ik}) = [\text{sign}(s_{ik,1}), \text{sign}(s_{ik,2})]^T$.

Define $V_0 = \frac{1}{2} s_{ik}^T s_{ik}$. Differentiating V_0 with respect to time t and substituting (2.5) and (2.7) into the derivative yields

$$\dot{V}_0 = \mathbf{s}_{ik}^T \dot{\mathbf{s}}_{ik} = \mathbf{s}_{ik}^T [-\tilde{\kappa}_{ik} \text{sign}(\mathbf{s}_{ik}) - \tilde{\eta}_{ik} \mathbf{s}_{ik} + \tilde{\mathbf{K}}_{ik} \mathbf{d}_{ik}] \tag{2.8}$$

Pick up $\tilde{\kappa}_{ik} > \|\tilde{\mathrm{K}}_{ik}\|_{\infty} d_{ik}^*$. There exist $V_0 \geq 0$ and $\dot{V}_0 < 0$. In the sense of Lyapunov, the formation-control system is of asymptotic stability. However, such a result is only available in theory. It is not available in reality because $\tilde{\kappa}_{ik}$ is unknown.

From another perspective, a conservative boundary of the formation uncertainties can be assigned in order to guarantee the formation stability, meaning the assigned boundary is larger or much larger than the actual boundary. A potential hazard of such a conservative design is that the conservative boundary can induce the chattering phenomenon and deteriorate the control performance.

Design of Sliding Mode Controller Based on NDOB

To attack the issue of the unknown d_{ik}^*, the NDOB technique is adopted to resist the adverse effects of the formation uncertainties. The following assumption about the formation uncertainties is taken into account.

Assumption 2: Compared with the dynamic characteristics of a NDOB, the change rate of the formation uncertainties in (2.3) is very slow, i.e., $\dot{\mathrm{d}}_{ik} \approx 0_{2\times1}$, where $0_{2\times1} = [0\ 0]^T$.

(2.1) and (2.4) are equivalent in mathematics but different in their forms. In order to facilitate the NDOB design, (2.4) is employed and the NDOB is designed by

$$\begin{cases} \dot{\mathbf{p}}_{ik} = -\mathbf{L}_{ik} \mathbf{B}_{ik,2} \mathbf{p}_{ik} - \mathbf{L}_{ik} (\mathbf{B}_{ik,2} \mathbf{L}_{ik} \mathbf{x}_{ik} + \mathbf{A}_{ik} \mathbf{x}_{ik} + \mathbf{B}_{ik,1} \mathbf{u}_k) \\ \hat{\mathbf{d}}_{ik} = \mathbf{p}_{ik} + \mathbf{L}_{ik} \mathbf{x}_{ik} \end{cases} \tag{2.9}$$

where $\mathrm{p}_{ik} \in \mathfrak{R}^{2\times1}$ and $\hat{\mathrm{d}}_{ik} \in \mathfrak{R}^{2\times1}$ are the internal state vector of the observer and the estimate of d_{ik}, respectively; $\mathrm{L}_{ik} \in \mathfrak{R}^{2\times4}$ is the observer gain matrix to be designed.

Define the estimate error vector

$$\mathbf{e}_{dik} = \mathbf{d}_{ik} - \hat{\mathbf{d}}_{ik} \tag{2.10}$$

Differentiate e_{dik} with respect to t and consider Assumptions 1 and 2. Then, the dynamics of e_{dik} can be described by

$$\begin{aligned}
\dot{\mathbf{e}}_{dik} &= \dot{\mathbf{d}}_{ik} - \dot{\hat{\mathbf{d}}}_{ik} \\
&\simeq -\dot{\mathbf{p}}_{ik} - \mathbf{L}_{ik}\dot{\mathbf{x}}_{ik} \\
&= \mathbf{L}_{ik}\mathbf{B}_{ik,2}\mathbf{p}_{ik} + \mathbf{L}_{ik}(\mathbf{B}_{ik,2}\mathbf{L}_{ik}\mathbf{x}_{ik} + \mathbf{A}_{ik}\mathbf{x}_{ik} + \mathbf{B}_{ik,1}\mathbf{u}_k) \\
&\quad -\mathbf{L}_{ik}(\mathbf{A}_{ik}\mathbf{x}_{ik} + \mathbf{B}_{ik,1}\mathbf{u}_k + \mathbf{B}_{ik,2}\mathbf{d}_{ik}) \\
&= \mathbf{L}_{ik}\mathbf{B}_{ik,2}(\hat{\mathbf{d}}_{ik} - \mathbf{L}_{ik}\mathbf{x}_{ik}) + \mathbf{L}_{ik}(\mathbf{B}_{ik,2}\mathbf{L}_{ik}\mathbf{x}_{ik} + \mathbf{A}_{ik}\mathbf{x}_{ik} \\
&\quad +\mathbf{B}_{ik,1}\mathbf{u}_k) - \mathbf{L}_{ik}(\mathbf{A}_{ik}\mathbf{x}_{ik} + \mathbf{B}_{ik,1}\mathbf{u}_k + \mathbf{B}_{ik,2}\mathbf{d}_{ik}) \\
&= \mathbf{L}_{ik}\mathbf{B}_{ik,2}(\hat{\mathbf{d}}_{ik} - \mathbf{d}_{ik}) \\
&= -\mathbf{L}_{ik}\mathbf{B}_{ik,2}\mathbf{e}_{dik}
\end{aligned} \tag{2.11}$$

Solving (2.11) yields

$$\mathbf{e}_{dik} = \exp(-\mathbf{H}_{ik}t)\mathbf{e}_{dik}(0) \tag{2.12}$$

here $H_{ik} = L_{ik} B_{ik,2}$.

Select L_{ik} such that all eigenvalues of H_{ik} are positive constants. This fact indicates the disturbance estimate error e_{dik} is exponentially convergent to $0_{2\times1}$ as $t\to\infty$.

Consider the formation dynamics (2.3) and the observer (2.9). Re-define the sliding surface vector as

$$\mathbf{s}_{ik} = \mathbf{J}_{ik}\mathbf{e}_{ik} + \mathbf{K}_{ik}\dot{\mathbf{e}}_{ik} + \hat{\mathbf{d}}_{ik} \tag{2.13}$$

here J_{ik} and K_{ik} are 2×2 positive-definite matrixes to be designed, written by

$$\mathbf{J}_{ik} = \begin{bmatrix} J_1 & 0 \\ 0 & J_2 \end{bmatrix} \text{ and } \mathbf{K}_{ik} = \begin{bmatrix} K_1 & 0 \\ 0 & K_2 \end{bmatrix} \tag{2.14}$$

Furthermore, the 1st-order SMC law based on NDOB can be determined by

$$\mathbf{u}_k = -(\mathbf{K}_{ik}\mathbf{G}_{ik})^{-1}[\kappa_{ik}\text{sign}(\mathbf{s}_{ik}) + \eta_{ik}\mathbf{s}_{ik} + \mathbf{J}_{ik}\dot{\mathbf{e}}_{ik} + \mathbf{K}_{ik}\hat{\mathbf{d}}_{ik}] \tag{2.15}$$

where κ_{ik} and η_{ik} are positive constants to be designed.

The system structure is presented in Figure 2.1. From Figure 2.1, l_{ik}, ψ_{ik} and their first derivatives are located at the feedback channel. They feed the NDOB. Furthermore, l_{ik} and ψ_{ik} construct the estimate error vector e_{ik}. e_{ik}, the derivative of e_{ik} and the estimate of d_{ik} feed the sliding surface vector s_{ik}. The 1st-order SMC law based on NDOB employs s_{ik}, the derivative of e_{ik} and the estimate of d_{ik} to generate the final control input vector u_k. Finally, u_k is applied to the follower k in order to maintain the desired distance and the desired relative angle during formation maneuvers.

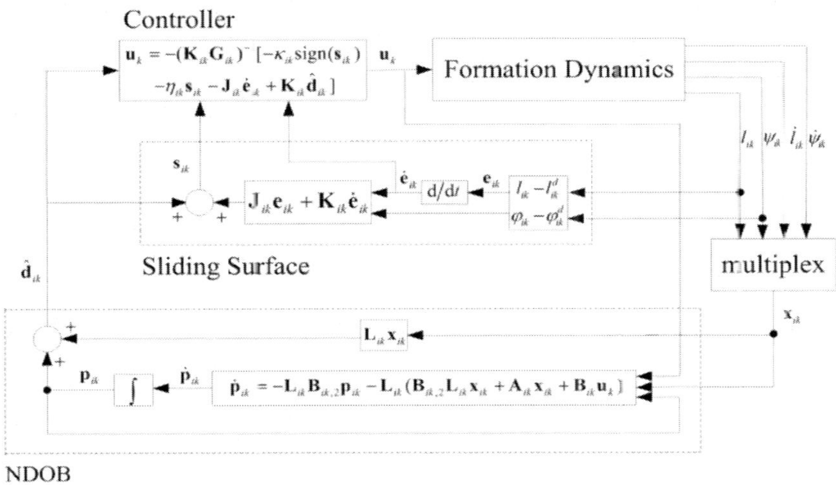

Figure 2.1. Schematic diagram of the designed control structure.

STABILITY ANALYSIS

Assumption 3: The estimate error vector e_{dik} is bounded by $\|\mathrm{e}_{dik}\|_\infty \le e^*_{dik}$, here $e^*_{dik} > 0$ is unknown.

Theorem 1: Focus on this leader-follower pair; take Assumptions 1, 2 and 3 into accounts, consider the formation dynamics (2.1), design the observer (2.9), define the sliding surface vector (2.13) and adopt the 1st-order SMC law based on NDOB (2.15). Then, the closed-loop formation system is asymptotically stable in the presence of uncertainties if $\kappa_{ik} > \|\mathrm{K}_{ik} + \mathrm{H}_{ik}\|_\infty e^*_{dik}$

Proof: Differentiate s_{ik} with respect to t. Substituting (7) into the derivative of s_{ik} yields

$$\dot{\mathbf{s}}_{ik} = \mathbf{J}_{ik}\dot{\mathbf{e}}_{ik} + \mathbf{K}_{ik}(\mathbf{G}_{ik}\mathbf{u}_k + \mathbf{d}_{ik}) + \dot{\hat{\mathbf{d}}}_{ik} \tag{2.16}$$

Substitute (2.15) into (2.16). Then (2.16) becomes

$$\dot{\mathbf{s}}_{ik} = -\kappa_{ik}\,\mathrm{sign}(\mathbf{s}_{ik}) - \eta_{ik}\mathbf{s}_{ik} + \mathbf{K}_{ik}(\mathbf{d}_{ik} - \hat{\mathbf{d}}_{ik}) + \dot{\hat{\mathbf{d}}}_{ik} \tag{2.17}$$

Furthermore, (2.18) can be derived from (2.11).

$$\dot{\hat{\mathbf{d}}}_{ik} = -\mathbf{H}_{ik}(\hat{\mathbf{d}}_{ik} - \mathbf{d}_{ik}) \tag{2.18}$$

Substitute (2.18) into (2.17). Then, $\dot{\mathrm{s}}_{ik}$ has a form of

$$\dot{\mathbf{s}}_{ik} = -\kappa_{ik}\mathrm{sign}(\mathbf{s}_{ik}) - \eta_{ik}\mathbf{s}_{ik} + (\mathbf{K}_{ik} + \mathbf{H}_{ik})(\mathbf{d}_{ik} - \hat{\mathbf{d}}_{ik}) \tag{2.19}$$

Select the following Lyapunov candidate function

$$V(t) = \|\mathbf{s}_{ik}\|_2 \tag{2.20}$$

here $\|\bullet\|_2$ denotes 2-norm.

Differentiate V with respect to time t. The derivative of V is obtained as

$$\dot{V}(t)=\frac{\mathbf{s}_{ik}^T\dot{\mathbf{s}}_{ik}}{\|\mathbf{s}_{ik}\|_2} \tag{2.21}$$

Substitute (2.19) into (2.21). Then, (2.22) can be gotten.

$$\begin{aligned}
\dot{V} &= \frac{\mathbf{s}_{ik}^T}{\|\mathbf{s}_{ik}\|_2}[-\kappa_{ik}\,\mathrm{sign}(\mathbf{s}_{ik})-\eta_{ik}\mathbf{s}_{ik}+(\mathbf{K}_{ik}+\mathbf{H}_{ik})(\mathbf{d}_{ik}-\hat{\mathbf{d}}_{ik})] \\
&= -\kappa_{ik}\frac{\mathbf{s}_{ik}^T\,\mathrm{sign}(\mathbf{s}_{ik})}{\|s_{ik}\|_2}-\eta_{ik}\frac{\mathbf{s}_{ik}^T\mathbf{s}_{ik}}{\|s_{ik}\|_2}+\frac{\mathbf{s}_{ik}^T}{\|\mathbf{s}_{ik}\|_2}(\mathbf{K}_{ik}+\mathbf{H}_{ik})(\mathbf{d}_{ik}-\hat{\mathbf{d}}_{ik}) \\
&= -\kappa_{ik}\frac{\|\mathbf{s}_{ik}\|_1}{\|\mathbf{s}_{ik}\|_2}-\eta_{ik}\frac{\|\mathbf{s}_{ik}\|_2^2}{\|\mathbf{s}_{ik}\|_2}+\frac{\mathbf{s}_{ik}^T}{\|\mathbf{s}_{ik}\|_2}(\mathbf{K}_{ik}+\mathbf{H}_{ik})\mathbf{e}_{dik} \\
&\le -\kappa_{ik}\frac{\|\mathbf{s}_{ik}\|_1}{\|\mathbf{s}_{ik}\|_2}-\eta_{ik}\frac{\|\mathbf{s}_{ik}\|_2^2}{\|\mathbf{s}_{ik}\|_2}+\frac{\mathbf{s}_{ik}^T}{\|\mathbf{s}_{ik}\|_2}\begin{bmatrix}\|(\mathbf{K}_{ik}+\mathbf{H}_{ik})\mathbf{e}_{dik}\|_\infty \\ \|(\mathbf{K}_{ik}+\mathbf{H}_{ik})\mathbf{e}_{dik}\|_\infty\end{bmatrix} \\
&\le -\kappa_{ik}\frac{\|\mathbf{s}_{ik}\|_1}{\|\mathbf{s}_{ik}\|_2}-\eta_{ik}\frac{\|\mathbf{s}_{ik}\|_2^2}{\|\mathbf{s}_{ik}\|_2}+\frac{\|\mathbf{s}_{ik}\|_1}{\|\mathbf{s}_{ik}\|_2}\|(\mathbf{K}_{ik}+\mathbf{H}_{ik})\mathbf{e}_{dik}\|_\infty \\
&\le -\kappa_{ik}\frac{\|\mathbf{s}_{ik}\|_1}{\|\mathbf{s}_{ik}\|_2}-\eta_{ik}\frac{\|\mathbf{s}_{ik}\|_2^2}{\|\mathbf{s}_{ik}\|_2}+\frac{\|\mathbf{s}_{ik}\|_1}{\|\mathbf{s}_{ik}\|_2}\|(\mathbf{K}_{ik}+\mathbf{H}_{ik})\|_\infty e_{dik}^* \\
&= -\left[\kappa_{ik}-\|(\mathbf{K}_{ik}+\mathbf{H}_{ik})\|_\infty e_{dik}^*\right]\frac{\|\mathbf{s}_{ik}\|_1}{\|\mathbf{s}_{ik}\|_2}-\eta_{ik}\frac{\|\mathbf{s}_{ik}\|_2^2}{\|\mathbf{s}_{ik}\|_2}
\end{aligned} \tag{2.22}$$

where $\|\cdot\|_1$ denotes 1-norm.

Let $\kappa_{ik} > \|K_{ik}+H_{ik}\|_\infty e_{dik}^*$. Then, $V < 0$ exists in (2.22). Furthermore, $V \geq 0$ exists in (2.20). The closed-loop system is asymptotically stable in the sense of Lyapunov.

Since e_{dik}^* is unknown in Assumption 3, κ_{ik} has to be assigned as well. In order to guarantee the formation stability, such a value must be conservative. At first glance, the 1st-order SMC law based on NDOB (2.15) makes no difference from the sole 1st-order SMC law (2.7). In fact, they are completely different. From (2.18), e_{dik} could be exponentially convergent to $0_{2\times1}$ by selecting L_{ik} such that κ_{ik} could be very small. Even if κ_{ik} is assigned from the conservative perspective, its value may be not very large. On the other hand, $\tilde{\kappa}_{ik}$ in (2.7) is determined by d_{ik}^*. A larger $\tilde{\kappa}_{ik}$ has to be assigned to guarantee the formation stability because d_{ik}^* is completely unknown. Consequently, the 1st-order SMC method based on NDOB can substantially alleviate the chattering problem of the SMC methodology.

Simulation Results

To verify the feasibility and validity of the presented control scheme for the formation-control problem of multiple robots with uncertainties, some numerical simulations are carried out in this section. The control scheme will be applied to a multi-robot system. The multi-robot system contains three identical robots. The robots are 0.05m in radius, where one robot is entitled leader and the other two are followers.

According to the idea of the leader-following formation, the leader is equipped by the sliding mode controller to track its predefined trajectories. The two followers is equipped by the presented sliding mode controllers based on NDOB to maintain the desired distances and the desired relative angles with respect to the leader in spite of uncertainties.

The parameter variations of the robots are considered by

$$\mathbf{\Delta}_i = \mathbf{\Delta}_k = \begin{bmatrix} 0.3\,\text{rand}() - 0.15 & 0 \\ 0 & 0.3\,\text{rand}() - 0.15 \end{bmatrix} \tag{2.23}$$

where $i = 1$ means the leader and $k = 2, 3$ denotes the followers. rand() is a MATLAB command. The command can generate a uniformly distributed random number. Note that (2.23) means that, the variations in the dynamical process are 15% of the nominal parameters. Generally speaking, the 15% variation is large enough to cover the regular parameter perturbation because too much variation may result in the motor overload in practice.

P in (2.1) is related to π_n. Note the terms $\pi_{ix}-\pi_{kx}$ and $\pi_{iy}-\pi_{ky}$ exist in P. The two terms indicate that it is not representative to define $\pi_i = \pi_k$. Here the following functions are assigned to the leader and the followers, determined by

$$\begin{aligned} \pi_{ix} &= \pi_{iy} = \pi_{i\theta} = 0.5\sin(2\pi t) \\ \pi_{kx} &= \pi_{ky} = \pi_{k\theta} = 0.2\sin(\pi t) \end{aligned} \tag{2.24}$$

Concerning the presented control scheme, some parameters of the sliding mode controllers are determined by $J_{12} = J_{13} = diag\{2, 2\}$, $K_{12} = K_{13} = diag\{3, 3\}$, $\kappa_{12} = \kappa_{13} = 2$ and $\eta_{12} = \eta_{13} = 3$, respectively. Additionally, the gains of the observers is selected by

$$\mathbf{L}_{12} = \mathbf{L}_{13} = \begin{bmatrix} 0 & 2 & 0 & 0.1 \\ 0 & 2 & 0 & 1 \end{bmatrix} \tag{2.25}$$

String Formation Moving in a Circular Trajectory

The first formation task of the multi-robot system is to form up a string formation while moving in a circular trajectory. The initial posture vector of the leader is located at $q_1 = [0.5\text{m}\ 0\text{m}\ \pi/2\text{rad}]^T$. The initial posture vectors of the followers are set by $q_2 = [0.8\text{m}\ -0.4\text{m}\ 0\text{rad}]^T$ and $q_3 = [1\text{m}\ 0.5\text{m}\ \pi\text{rad}]^T$, respectively. According to these posture vectors, the initial state vectors of the followers with respect to the leader can be calculated by $x_{12}^0 = [0.5\text{m}\ 0\text{m/s}\ 3.2\pi/4\text{rad}\ 0\text{rad/s}]^T$ and $x_{13}^0 = [0.707\text{m}\ 0\text{m/s}\ 3\pi/4\text{rad}\ 0\text{rad/s}]^T$. The desired state vectors are set by $x_{12}^d = [0.13\text{m}\ 0\text{m/s}\ \pi/2\text{rad}\ 0\text{rad/s}]^T$ and $x_{13}^d = [0.26\text{m}\ 0\text{m/s}\ \pi/2\text{rad}\ 0\text{rad/s}]^T$. Considering the motor load, the acceleration and the angular acceleration of the followers are subject to $|a_2| = |a_3| \leq 0.5\pi\text{m/s}^2$ and $|\beta_2| = |\beta_3| \leq 15\text{rad/s}^2$, respectively. The desired linear and angular velocities of the leader are set by $v_1 = 0.5\text{m/s}$ and $w_1 = 1\text{rad/s}$.

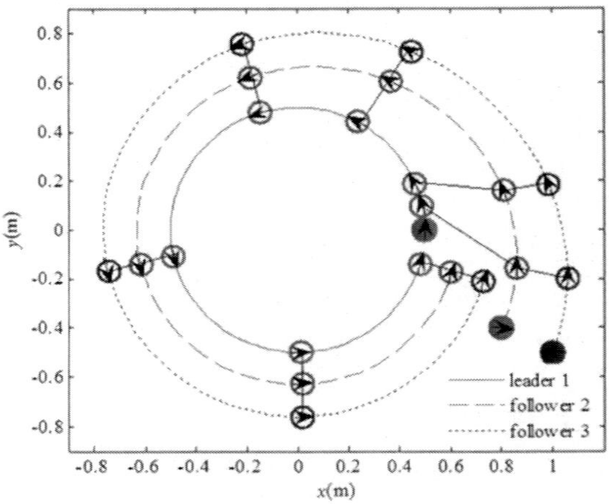

Figure 2.2. String formation while moving in a circular trajectory.

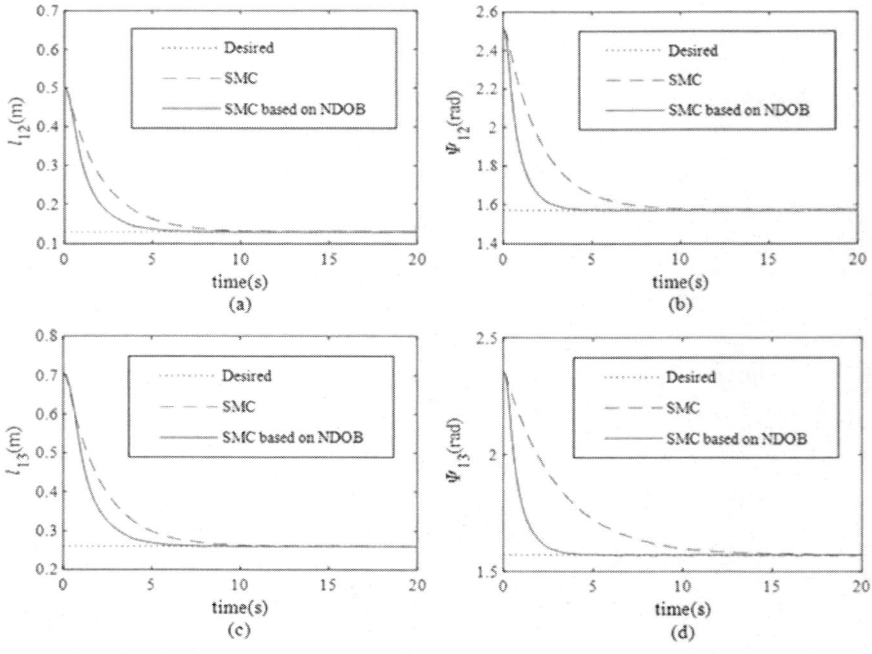

Figure 2.3. Distances and relative angles of the followers. (a). l_{12} of follower 2, (b). ψ_{12} of follower 2, (c). l_{13} of follower 3 and (d). ψ_{13} of follower 3.

Figure 2.2 demonstrates the robots forms up the string while moving in the circular trajectory. In Figure 2.2, the solid points denote the initial positions; the arrows indicate the orientation angles and the lines connect the robots together at the same moment. From Figure 2.2, the leader tracks its predefined circular trajectory and the followers maintain the desired distances and the desired angles to follow the leader. The formation task has been achieved.

Although Figure 2.2 illustrates the formation maneuver, it does not provide any insight into the control performance. Some comparisons among the SMC method based on NDOB and the sole SMC method are displayed in Figures 2.3-2.7. The sole SMC design has been presented, where the parameters of the sliding mode controllers are kept uncharged from the presented control scheme except $\tilde{\kappa}_{12} = \tilde{\kappa}_{13} = 6$ from the aspect of the system stability.

In Figure 2.3a and b, the curves denote the distance and relative angle of follower 2. In Figure 2.3c and d, the curves indicate the distance and relative angle of follower 3. Compared with the results by the sole SMC method, the performance of the SMC method based on NDOB in Figure 2.3 is fastest and smoothest without doubt although the other two methods have achieved the formation maneuver despite uncertainties as well.

Figure 2.4 demonstrates the control inputs of follower 2 by the two control methods, where the control inputs by the SMC method based on NDOB are shown in Figure 2.4a and b; the control inputs by the sole SMC method are shown in Figure 2.4c and d. In Figure 2.4, the curves of the control inputs by the sole SMC method severely chatter because a conservative boundary of the formation uncertainties is assigned in order to guarantee the formation stability. Here, the boundary is as large as 6 in the simulation. Without doubt, the curves of the control inputs by the SMC method based on NDOB are superior. Although the boundary of the estimate error vector is unknown according to the presented control scheme, a conservative boundary of the estimate error vector needs to be assigned in advance. However, the conservative boundary is related to the estimate error vector. The aforementioned analysis points out the estimate error vector can be exponentially convergent. The boundary of the estimate error vector is much smaller than the boundary of the formation uncertainties. As a result, the presented control scheme can alleviate the chattering problem so that the performance of the control inputs can be improved. Figure 2.5 demonstrates the control inputs of follower 3 by the two control methods. Like the results in Figure 2.4, some similar results can be drawn from Figure 2.5 as well.

Figure 2.6 illustrates the sliding surface vectors and their elements. As proven in Theorem 1, the presented control scheme can guarantee the formation stability of the multi-robot system. Furthermore, the sliding modes of the followers are reachable in finite time. In Figure 2.6, the formation-control system spends about 3 seconds in length to enter the sliding modes. Although the trajectories in Figure 2.6c and f may deviate from the sliding surfaces in the presence of uncertainties, the presented control scheme can make the formation control system asymptotically stable.

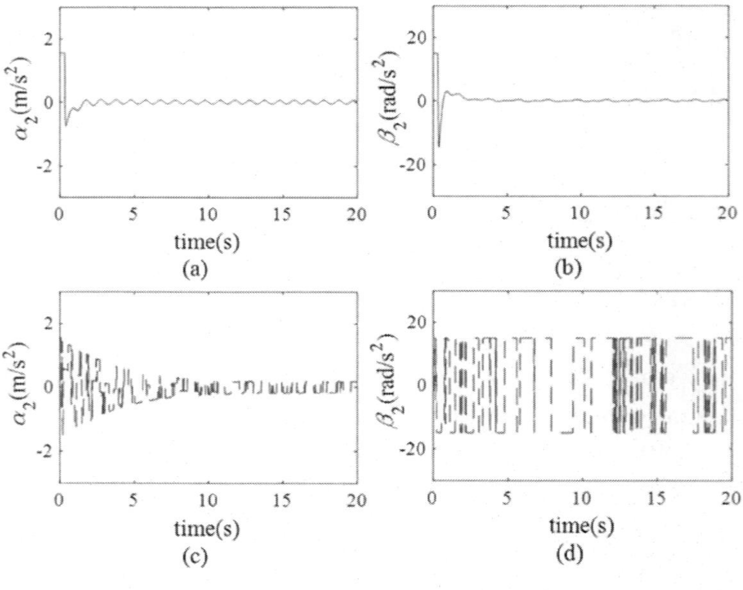

Figure 2.4. Accelerations and angular accelerations of follower 2 by the three control methods. (a). α_2 by the presented method, (b). β_2 by the presented method, (c). α_2 by the sole SMC method, and (d). β_2 by the sole SMC method.

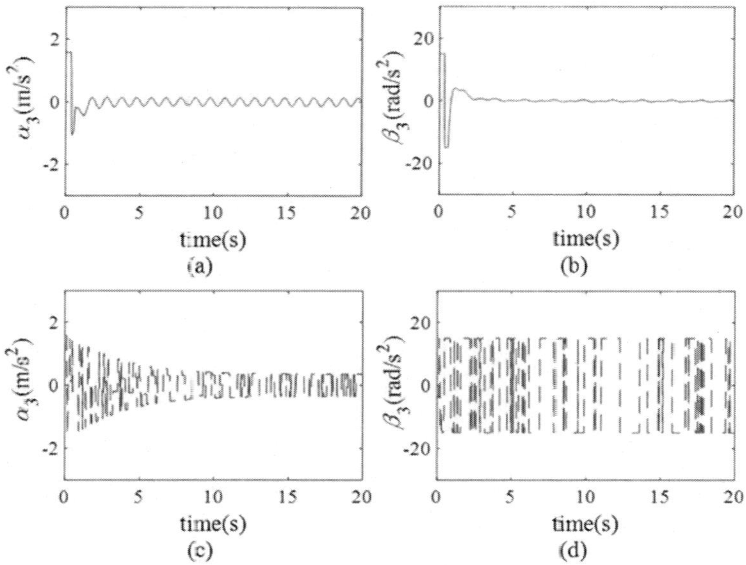

Figure 2.5. Accelerations and angular accelerations of follower 3 by the three control methods. (a). α_3 by the presented method, (b). β_3 by the presented method, (c). α_3 by the sole SMC method, and (d). β_3 by the sole SMC method.

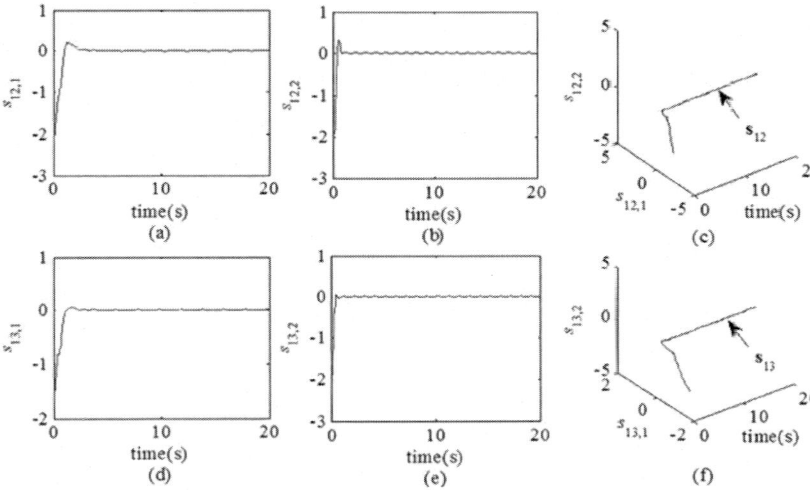

Figure 2.6. Sliding surfaces. (a). $s_{12,1}$, (b). $s_{12,2}$, (c) three-dimensional plot of $\mathbf{s}_{12}$, (d). $s_{13,1}$, (e). $s_{13,2}$ and (f). three-dimensional plot of $\mathbf{s}_{13}$.

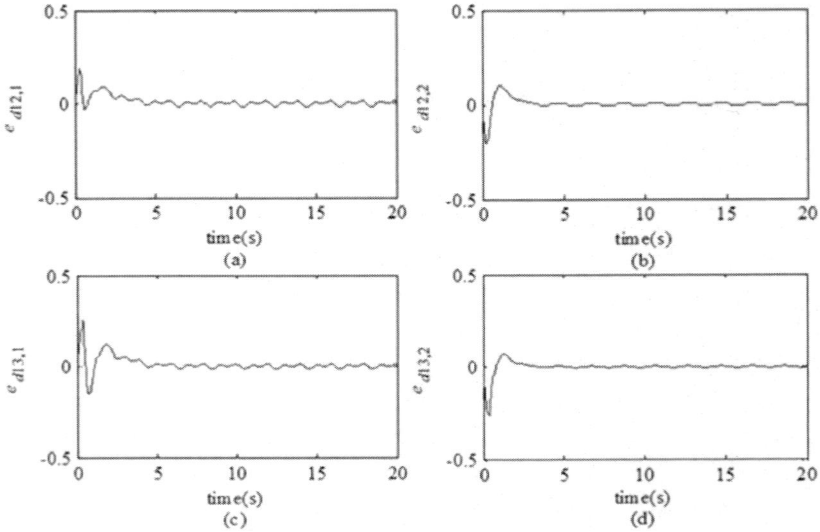

Figure 2.7 Elements of the estimate error vectors. (a). $e_{d12,1}$, (b). $e_{d12,2}$, (c). $e_{d13,1}$ and (d). $e_{d13,2}$.

Figure 2.7 displays the elements of the estimate error vectors, where the elements of e_{d12} are shown in Figure 2.7a and b and the elements of e_{d13} are shown in Figure 2.7c and d. From Figure 2.7, both the estimate error vectors are convergent to $0_{2\times1}$. The maximums of the two vectors in Figure 2.7 are less than 0.5. On the other hand, the predefined boundary of e_{dik} is 2 such that the boundary is conservative enough to guarantee the formation stability in theory.

Transformation Moving in a Straight Trajectory

This formation task of the multi-robot system is compound. Firstly, the robots form up a string while moving a straight trajectory. Then, they transform the string into a triangle. The maneuver is illustrated in Figure 2.8. From Figure 2.8, the solid points denote the initial positions of the robots; the arrows mean the orientation angles of the robots and the line connect the robots together at the same moment. In Figure 2.8, the initial postures are located at $q_1 = [0\text{m}\ 0\text{m}\ 0\pi\text{rad}]^T$, $q_2 = [0\text{m}\ 0.4\text{m}\ 0\text{rad}]^T$ and $q_3 = [0\text{m}\ -0.3\text{m}\ 0\text{rad}]^T$, respectively. According to these initial postures, $x_{12}^0 = [0.4\text{m}\ 0\text{m/s}\ \pi/2\text{rad}\ 0\text{rad/s}]^T$ and $x_{13}^0 = [0.3\text{m}\ 0\text{m/s}\ -\pi/2\text{rad}\ 0\text{rad/s}]^T$ can be calculated. All the parameters of the presented control scheme are kept unchanged from the aforementioned two formation tasks.

For the duration $0 < t \leq 8$s, the desired states of the followers are set by $x_{12}^d = [0.2\text{m}\ 0\text{m/s}\ \pi/2\text{rad}\ 0\text{rad/s}]^T$ and $x_{13}^d = [0.15\text{m}\ 0\text{m/s}\ -\pi/2\text{rad}\ 0\text{rad/s}]^T$ to form up the string formation. For the subsequent duration t>8s, the desired states of the followers are set by $x_{12}^d = [0.5\text{m}\ 0\text{m/s}\ 2\pi/3\text{rad}\ 0\text{rad/s}]^T$ and $x_{13}^d = [0.5\text{m}\ 0\text{m/s}\ -2\pi/3\text{rad}\ 0\text{rad/s}]^T$. Concerning the leader, its desired linear and angular velocities are set by $v_1 = 0.5$m/s and $w_1 = 0$rad/s.

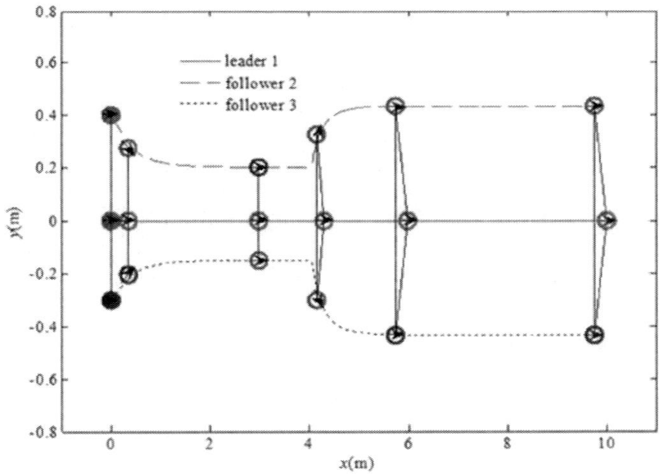

Figure 2.8. Transformation of the multi-robot system while moving in a straight.

CONCLUSION

Concerning the formation-control problem of multiple robots, this article has proposed a control scheme. The control scheme is combined the SMC method and the NDOB technique. Chained to uncertainties of the robots, their leader-following formation dynamics contain the formation uncertainties. The formation uncertainties not only challenge the formation control design, but also can they deteriorate the formation performance. Assuming that the formation uncertainties have an unknown boundary, this article adopts a NDOB to estimate the formation uncertainties. A sliding surface vector is designed by means of the NDOB estimate. Thereafter, the SMC law is developed. In the sense of Lyapunov, the formation stability in the presence of uncertainties is proven. Compared with the results by the sole SMC method, some simulation results are presented to exhibit that the proposed control scheme is preferable.

REFERENCES

[1] Wehner, M., Truby, R. L., Fitzgerald, D. J., Mosadegh, B., Whitesides, G. M., Lewis, J. A., Wood, R. J. (2016). An integrated design and fabrication strategy for entirely soft, autonomous robots. *Nature*, 536(7617), 451-455.

[2] Ingrand, F., Ghallab, M. (2017). Deliberation for autonomous robots: A survey. *Artificial Intelligence*, 247, 10-44.

[3] Issa, B. A., Rashid, A. T. (2019). A Survey of Multi-mobile Robot Formation Control. *International Journal of Computer Applications*, 975, 8887.

[4] Das, A. K., Fierro, R., Kumar, V., Ostrowski, J. P., Spletzer, J., Taylor, C. J. (2002). A vision-based formation control framework. *IEEE transactions on robotics and automation*, 18(5), 813-825.

[5] Chen, J., Sun, D., Yang, J., Chen, H. (2010). Leader-follower formation control of multiple non-holonomic mobile robots incorporating a receding-horizon scheme. *The International Journal of Robotics Research*, 29(6), 727-747.

[6] Li, X., Xiao, J. (2005). Robot formation control in leader-follower motion using direct Lyapunov method. *International Journal of Intelligent Control and Systems*, 10(3), 244-250.

[7] Consolini, L., Morbidi, F., Prattichizzo, D., Tosques, M. (2008). Leader–follower formation control of nonholonomic mobile robots with input constraints. *Automatica*, 44(5), 1343-1349.

[8] Li, Z., Yuan, Y., Ke, F., He, W., Su, C. Y. (2019). Robust Vision-Based Tube Model Predictive Control of Multiple Mobile Robots for Leader–Follower Formation. *IEEE Transactions on Industrial Electronics*, 67(4), 3096-3106.

[9] Ren, W., Beard, R. W., Atkins, E. M. (2005, June). A survey of consensus problems in multi-agent coordination. In *Proceedings of the 2005 American Control Conference*, 2005. (pp. 1859-1864). IEEE.

[10] Zhu, D. Q., Yan, M. Z. (2010). Survey on technology of mobile robot path planning. *Control and Decision*, 25(7), 961-967.

[11] Sandoval-Rodriguez, R., Abdallah, C. T., Hokayem, P. F., Schamiloglu, E., & Byrne, R. H. (2003, December). Robust mobile robotic formation control using internet-like protocols. In *Proceedings of the 42nd IEEE International Conference on Decision and Control* (Vol. 5, pp. 5109-5112). IEEE.

[12] Xiao, H., Li, Z., Chen, C. P. (2016). Formation control of leader–follower mobile robots' systems using model predictive control based on neural-dynamic optimization. *IEEE Transactions on Industrial Electronics*, 63(9), 5752-5762.

[13] Wang, W., Huang, J., Wen, C., Fan, H. (2014). Distributed adaptive control for consensus tracking with application to formation control of nonholonomic mobile robots. *Automatica*, 50(4), 1254-1263.

[14] Lawton, J. R., Beard, R. W., Young, B. J. (2003). A decentralized approach to formation maneuvers. *IEEE transactions on robotics and automation*, 19(6), 933-941.

[15] Chang, Y. H., Chang, C. W., Chen, C. L., Tao, C. W. (2011). Fuzzy sliding-mode formation control for multirobot systems: design and implementation. *IEEE Transactions on Systems, Man, and Cybernetics, Part B−Cybernetics*, 42(2), 444-457.

[16] Young, K. D., Utkin, V. I., Ozguner, U. (1999). A control engineer's guide to sliding mode control. *IEEE transactions on control systems technology*, 7(3), 328-342.

[17] Utkin, V., Guldner, J., Shi, J. (2017). *Sliding mode control in electro-mechanical systems*. CRC press.

[18] Perruquetti, W., Barbot, J. P. (2002). *Sliding mode control in engineering*. CRC press.

[19] Yang, L., Yang, J. (2011). Nonsingular fast terminal sliding-mode control for nonlinear dynamical systems. *International Journal of Robust and Nonlinear Control*, 21(16), 1865-1879.

[20] Nair, R. R., Behera, L., Kumar, V., Jamshidi, M. (2014). Multisatellite formation control for remote sensing applications using artificial potential field and adaptive fuzzy sliding mode control. *IEEE Systems Journal*, 9(2), 508-518.

[21] Cao, L., Chen, X. (2015). Input–output linearization minimum sliding-mode error feedback control for spacecraft formation with large perturbations. *Proceedings of the Institution of Mechanical Engineers, Part G: Journal of Aerospace Engineering*, 229(2), 352-368.

[22] Defoort, M., Floquet, T., Kokosy, A., Perruquetti, W. (2008). Sliding-mode formation control for cooperative autonomous mobile robots. *IEEE Transactions on Industrial Electronics*, 55(11), 3944-3953.

[23] Chen, W. H. (2004). Disturbance observer based control for nonlinear systems. *IEEE/ASME transactions on mechatronics*, 9(4), 706-710.

[24] Mohammadi, A., Tavakoli, M., Marquez, H. J., Hashemzadeh, F. (2013). Nonlinear disturbance observer design for robotic manipulators. *Control Engineering Practice*, 21(3), 253-267.

[25] Yang, J., Li, S., Sun, C., Guo, L. (2013). Nonlinear-disturbance-observer-based robust flight control for airbreathing hypersonic vehicles. *IEEE Transactions on Aerospace and Electronic Systems*, 49(2), 1263-1275.

[26] Nikoobin, A., Haghighi, R. (2009). Lyapunov-based nonlinear disturbance observer for serial n-link robot manipulators. *Journal of Intelligent and Robotic Systems*, 55(2-3), 135-153.

[27] Wang, C., Li, X., Guo, L., Li, Y. W. (2014). A nonlinear-disturbance-observer-based DC-bus voltage control for a hybrid AC/DC microgrid. *IEEE Transactions on Power Electronics*, 29(11), 6162-6177.

[28] Qian, D., Tong, S., & Li, C. (2016). Leader-Following Formation Control of Multiple Robots with Uncertainties through Sliding Mode and Nonlinear Disturbance Observer. *ETRI Journal*, 38(5), 1008-1018.

In: A Closer Look at Formation Control ISBN: 978-1-53618-177-7
Editor: Dianwei Qian

Chapter 3

COORDINATED FORMATION VIA AN ADAPTIVE-GAIN 2ND-ORDER SLIDING MODE METHOD

Dianwei Qian[*] and Jiarong Chen
School of Control and Computer Engineering,
North China Electric Power University, Beijing, China

ABSTRACT

This chapter presents a 2nd-order sliding mode control method for the formation problem of multi-robot systems. The multiple robots contain plenty of uncertainties and disturbances. Such a control method is with two adaptive gains that can contribute to the robustness and improve the response of the formation maneuvers despite these uncertainties and disturbances. Based on the leader-follower frame, this control method is investigated. The closed-loop formation stability is theoretically guaranteed in the sense of Lyapunov. From the aspect of practice, the control method is carried out by a multi-robot system to achieve some desired formation patterns. Some numerical results are demonstrated to verify the feasibility of the control method. Some comparisons are also

[*] Corresponding Author's Email: dianwei.qian@ncepu.edu.cn.

illustrated to support the superiority and effectiveness of the designed sliding mode control method.

Keywords: adaptive control, 2nd-order sliding mode control, robotics, system stability

INTRODUCTION

With the coming of the artificial intelligence era, multi-robot systems are becoming more attractive and more significant [1, 2]. It is one of the emerging and popular solutions in various fields, such as industry, agriculture and aviation, etc. Concerning multi-robot systems, the control problem of formation maneuvers is one of the important topics [3, 4]. In many cases, the robots in a multi-robot system have to form up some formation patterns in order to accomplish a given task in collaborative projects, military reconnaissance and hazardous situations. For the purpose of organizing and managing the robots, a coordinated control scheme needs to be pre-defined in the multi-robot system [5−7].

In this chapter, the leader-follower scheme is taken into consideration because the scheme can accomplish a given formation task with the guaranteed closed-loop stability [8, 9]. Although it suffers from the drawback of a 'single point of failure', the leader-follower scheme is of merit for the formation maneuvers of small-and-medium-scale multi-robot systems [10, 11]. This chapter does not focus on how to design a coordinated control scheme, but it concentrates on the formation control design. Thus, the leader-follower scheme is directly adopted. In addition, the small-scale multi-robot systems are taken into account for the convenience of the control design [12].

Inherently, the formation model of a multi-robot system contains nonlinearities and motion couples that challenge the formation control design very much [13−15]. In reality, a multi-robot system is inevitably subjected to some uncertainties and disturbances, including but not limited to modeling errors, inertia-and-mass variations, signal-transmission delays, unpredictable obstacles, etc [16]. These uncertainties and disturbances challenge the formation control very much. To deal with the control problem of formation maneuvers, many control methods have been reported, i.e., model predict control [17], adaptive control [18], interval type-2 fuzzy control [19], adaptive dynamic programming [20], etc.

The methodology of sliding mode control is a synthetic tool [21, 22]. Some sliding-mode-based control approaches have been investigated for the control problem. Some studies have been reported, such as the first-order sliding mode control [23], the integral sliding mode [24, 25], the derivative and integral terminal sliding mode control [26], the terminal sliding mode control [27] and so on. As far as the methodology of sliding mode control is concerned, the invariance is its most attractive property, which means that a sliding-mode-based control system is completely and thoroughly robust against the uncertainties and disturbances entering the control system by the control channel. Dialectically, the methodology is criticized for its chattering phenomenon as well. Many sliding-mode-based control ideas focus on decreasing even eliminating the chattering phenomenon, where the super-twisting sliding mode control is such a kind of the 2nd-order sliding mode control methods.

The super-twisting sliding mode control becomes popular because this control technique only needs the information of a sliding mode variable and gets rid of the dependence on the time derivative of this sliding mode variable [28, 29]. The control technique can effectively force the sliding mode variable and its time derivative to the origin in finite time despite the existence of the bounded disturbances and uncertainties on the assumption that the boundary is known. Unfortunately, this boundary can hardly be known with regard to the formation maneuvers of multi-robot systems. One can overestimate this boundary from the aspect of the closed-loop stability [30, 31]. But the overestimate definitely enlarges the necessary control gain of the super-twisting sliding mode control [32, 33]. In order to deal with the issue, the gain adaptation algorithm is taken into account. The integration of the gain adaption algorithm and the super-twisting sliding mode control can benefit the formation maneuvers of uncertain multi-robot systems with the unknown boundary. However, the design is just presented for the single-input-single-output systems and it cannot be directly extended to a multiple-input-multiple-output system as the formation maneuvers of multi-robots. In this chapter, we focus on this field, work at the issue and seek to its solution. The purpose is to investigate an adaptive-gain super-twisting sliding mode control design for the formation maneuvers of multi-robot systems.

The remainder is presented as follows. Section 2 introduces the modeling of a single mobile robot as well as the modeling of a leader-follower pair of multi-robot systems. Section 3 presents the design of the adaptive-gain super-twisting sliding mode control and adopts the Lyapunov theory to analyze the closed-loop system stability. In Section 4, we implement the adaptive-gain super-twisting sliding mode control on a multi-robot system platform. Some numerical results and comparisons are illustrated in Section 4. Finally, conclusions are drawn in Section 5

FORMATION MODELING

Recall Chapter 1, the mathematical model (1.8) can be described by

$$\ddot{\mathbf{h}}_{ik} = \mathbf{G}(\mathbf{I}_2 + \Delta_k)\mathbf{u}_k + \mathbf{L}(\mathbf{I}_2 + \Delta_i)\mathbf{u}_i + \tilde{\mathbf{F}} + \tilde{\mathbf{P}} \tag{3.1}$$

Define a vector $\mathbf{x}_{ik} = [x_1\ x_2\ x_3\ x_4]^T$. Let $x_1 = l_{ik}$, $x_2 = \dot{l}_{ik}$, $x_3 = \psi_{ik}$ and $x_4 = \dot{\psi}_{ik}$. According to the formation objective, the relative distance l_{ik} and the relative bearing angle ψ_{ik} are determined as the formation control output. Then, the dynamics of formation maneuvers of the multi-robot system can have the form of (3.2) in light of the leader-follower scheme.

$$\begin{aligned} \dot{\mathbf{x}}_{ik} &= \mathbf{f}(\mathbf{x}_{ik}, \mathbf{d}_{ik}) + \mathbf{g}(\mathbf{x}_{ik}, \mathbf{\Delta}_k)\mathbf{u}_k \\ \mathbf{y}_{ik} &= \mathbf{h}(\mathbf{x}_{ik}) \end{aligned} \tag{3.2}$$

here $\mathbf{x}_{ik}$ is the system state vector and $\mathbf{y}_{ik}$ is the system output vector. Further,

$$\begin{aligned}\mathbf{f}(\mathbf{x}_{ik}, \mathbf{d}_{ik}) &= \mathbf{A}_{ik}\mathbf{x}_{ik} + \mathbf{B}_{ik,2}\mathbf{d}_{ik} \\ \mathbf{g}(\mathbf{x}_{ik}, \boldsymbol{\Delta}_k) &= \mathbf{B}_{ik,1} + \mathbf{B}_{ik,1}\boldsymbol{\Delta}_k\end{aligned} \tag{3.3}$$

$\mathbf{A}_{ik}$, $\mathbf{B}_{ik,1}$, $\mathbf{B}_{ik,2}$ and $\mathbf{h}(\mathbf{x}_{ik})$ are depicted by

$$\mathbf{A}_{ik} = \begin{bmatrix} 0 & 1 & 0 & 0 \\ 0 & 0 & 0 & 0 \\ 0 & 0 & 0 & 1 \\ 0 & 0 & 0 & 0 \end{bmatrix}, \quad \mathbf{B}_{ik,2} = \begin{bmatrix} 0 & 0 \\ 1 & 0 \\ 0 & 0 \\ 0 & 1 \end{bmatrix}, \quad \mathbf{B}_{ik,1} = \begin{bmatrix} 0 & 0 \\ \cos\varphi_{ik} & r\sin\varphi_{ik} \\ 0 & 0 \\ \dfrac{-\sin\varphi_{ik}}{l_{ik}} & \dfrac{r\cos\varphi_{ik}}{l_{ik}} \end{bmatrix}$$

$$\text{and } \mathbf{h}(\mathbf{x}_{ik}) = \begin{bmatrix} x_1 \\ x_3 \end{bmatrix} \tag{3.4}$$

where $\varphi_{ik} = \psi_{ik} + \theta_{ik}$. $\mathbf{d}_{ik}$ is the lumped term of all the uncertainties and disturbances in the leader-follower pair.

$$\mathbf{d}_{ik} = \mathbf{L}_{ik}(\mathbf{I}_2 + \boldsymbol{\Delta}_i)\mathbf{u}_i + \tilde{\mathbf{F}} + \tilde{\mathbf{P}} \tag{3.5}$$

In (3.5), we have

$$\mathbf{L}_{ik} = \begin{bmatrix} 0 & 0 \\ -\cos\psi_{ik} & 0 \\ 0 & 0 \\ \dfrac{\sin\psi_{ik}}{l_{ik}} & -1 \end{bmatrix}, \quad \tilde{\mathbf{F}} = \begin{bmatrix} 0 \\ F_1 \\ 0 \\ F_2 \end{bmatrix} \text{ and } \tilde{\mathbf{P}} = \begin{bmatrix} 0 \\ P_1 \\ 0 \\ P_2 \end{bmatrix} \tag{3.6}$$

Here $\mathbf{I}_2$ is a 2×2 identity matrix and F_1, F_2, P_1 and P_2 are written by

$$F_1 = (\dot{\psi}_{ik})^2 l_{ik} + 2\dot{\psi}_{ik}\dot{\theta}_i l_{ik} + (\dot{\theta}_i)^2 l_{ik} \\ -r\cos\varphi_{ik}(\dot{\theta}_k)^2 - (\dot{y}_k\dot{\theta}_k - \dot{y}_i\dot{\theta}_i)\cos(\psi_{ik} + \dot{\theta}_i) - (\dot{x}_i\dot{\theta}_i - \dot{x}_k\dot{\theta}_k)\sin(\psi_{ik} + \theta_i)$$

$$F_2 = \frac{-(\dot{y}_k\dot{\varphi}_{ik} - \dot{\psi}_{ik}\dot{y}_i)\sin(\psi_{ik} + \theta_i) - r\dot{\theta}_k\dot{\varphi}_{ik}\sin\varphi_{ik}}{l_{ik}} \\ \frac{-(\dot{x}_k\dot{\varphi}_{ik} - \dot{\psi}_{ik}\dot{x}_i)\cos(\psi_{ik} + \theta_i) + \dot{l}_{ik}((\dot{y}_i - \dot{y}_k)\cos(\psi_{ik} + \theta_i)}{l_{ik}} \\ \frac{-(\dot{x}_i - \dot{x}_k)\sin(\psi_{ik} + \theta_i) - r\dot{\theta}_k\dot{\varphi}_{ik}\cos\varphi_{ik})}{l_{ik}}$$

$$P_1 = -(\pi_{ix} - \pi_{kx})\cos(\psi_{ik} + \theta_i) - (\pi_{iy} - \pi_{ky})s\,\mathrm{in}(\psi_{ik} + \theta_i) + r\pi_{k\theta}\sin\varphi_{ik}$$

$$P_2 = \frac{(\pi_{ix} - \pi_{kx})s\,\mathrm{in}(\psi_{ik} + \theta_i) - (\pi_{iy} - \pi_{ky})\cos(\psi_{ik} + \theta_i) + r\pi_{k\theta}\sin\varphi_{ik} - l_{ik}\pi_{i\theta}}{l_{ik}} \tag{3.7}$$

CLOSED-LOOP CONTROL DESIGN

Sliding Mode Design and Its Input-Output Dynamics

The super-twisting law is a powerful second-order sliding-mode technique. It can effectively deal with the controlled plant with a relative degree equal to one with respect to the control input. Theoretically, the technique can make the sliding mode variable and its time derivative convergent to the origin in spite of the uncertainties and disturbances. Thus, this technique can be considered as a solution for formation maneuvers of the multi-robot system.

In order to implement the control design of the leader-follower pair, the sliding mode vector $\mathbf{s}_{ik}$, that is, the sliding surfaces, has to be defined in advance.

$$\mathbf{s}_{ik} = \begin{bmatrix} s_{ik,1} \\ s_{ik,2} \end{bmatrix} = \mathbf{C}_1\left(\begin{bmatrix} l_{ik} \\ \psi_{ik} \end{bmatrix} - \begin{bmatrix} l_{ik}^d \\ \psi_{ik}^d \end{bmatrix}\right) + \mathbf{C}_2\left(\begin{bmatrix} \dot{l}_{ik} \\ \dot{\psi}_{ik} \end{bmatrix} - \begin{bmatrix} \dot{l}_{ik}^d \\ \dot{\psi}_{ik}^d \end{bmatrix}\right) \tag{3.8}$$

Here $\mathbf{C}_1$ and $\mathbf{C}_2$ are 2×2 constant matrices and they need to be pre-defined. l_{ik}^d and ψ_{ik}^d are the desired relative distance and the desired relative bearing angle of this leader-follower pair, respectively.

Differentiate the sliding mode vector $\mathbf{s}_{ik}$ in (3.8) with the respect to time, the input–output dynamics can be derived as follows.

$$\dot{\mathbf{s}}_{ik} = \frac{\partial \mathbf{s}_{ik}}{\partial t} + \frac{\partial \mathbf{s}_{ik}}{\partial \mathbf{x}_{ik}} \mathbf{f}(\mathbf{x}_{ik}, \mathbf{d}_{ik}) + \frac{\partial \mathbf{s}_{ik}}{\partial \mathbf{x}_{ik}} \mathbf{g}(\mathbf{x}_{ik}, \mathbf{\Delta}_k) \mathbf{u}_k \qquad (3.9)$$

From (3.8) and (3.9), we have

$$\frac{\partial \mathbf{s}_{ik}}{\partial \mathbf{u}_k} = 0 \text{ and } \frac{\partial \dot{\mathbf{s}}_{ik}}{\partial \mathbf{u}_k} = \frac{\partial \mathbf{s}_{ik}}{\partial \mathbf{x}_{ik}} \mathbf{g}(\mathbf{x}_{ik}, \mathbf{\Delta}_k) \neq \mathbf{O} \qquad (3.10)$$

Namely, the relative degree of $\mathbf{s}_{ik}$ with respect to $\mathbf{u}_k$ is equal to 1. Subsequently, the super-twisting-based sliding mode control design can be available for formation maneuvers of such a multi-robot system.

Let

$$\begin{aligned} \mathbf{a}(\mathbf{x}_{ik}, \mathbf{d}_{ik}, t) &= \frac{\partial \mathbf{s}_{ik}}{\partial t} + \frac{\partial \mathbf{s}_{ik}}{\partial \mathbf{x}_{ik}} \mathbf{f}(\mathbf{x}_{ik}, \mathbf{d}_{ik}) \\ \mathbf{b}(\mathbf{x}_{ik}, \mathbf{\Delta}_k, t) &= \frac{\partial \mathbf{s}_{ik}}{\partial \mathbf{x}_{ik}} \mathbf{g}(\mathbf{x}_{ik}, \mathbf{\Delta}_k) \end{aligned} \qquad (3.11)$$

Assumption 1: $\mathbf{b}(\mathbf{x}_{ik}, \mathbf{\Delta}_k, t)$ is a 2×2 matrix and contains both known and unknown parts, written by

$$\mathbf{b}(\mathbf{x}_{ik}, \mathbf{\Delta}_k, t) = \mathbf{b}_0(\mathbf{x}_{ik}, t) + \Delta\mathbf{b}(\mathbf{x}_{ik}, \mathbf{\Delta}_k, t) \qquad (3.12)$$

Here $\mathbf{b}_0(\mathbf{x}_{ik}, t)$ is a known positively-definite matrix, $\Delta\mathbf{b}_0(\mathbf{x}_{ik}, \mathbf{\Delta}_k, t)$ is bounded but unknown and the two parts of $\mathbf{b}(\mathbf{x}_{ik}, \mathbf{\Delta}_k, t)$ satisfy

$$\| \Delta \mathbf{b}(\mathbf{x}_{ik}, \mathbf{\Delta}_k, t)\mathbf{b}_0^{-1}(\mathbf{x}_{ik}, t) \|_2 \leq \gamma(\mathbf{x}_{ik}, t) < \gamma_1 < 1 \tag{3.13}$$

where γ_1 is a unknown constant.

Assumption 2: $\mathbf{a}(\mathbf{x}_{ik}, \mathbf{d}_{ik}, t)$ is a 2×1 vector and contains both known and unknown parts, depicted by

$$\mathbf{a}(\mathbf{x}_{ik}, \mathbf{d}_{ik}, t) = \mathbf{a}_1(\mathbf{x}_{ik}, t) + \mathbf{a}_2(\mathbf{x}_{ik}, \mathbf{d}_{ik}, t) \tag{3.14}$$

Here both $\mathbf{a}_1(\mathbf{x}_{ik}, t)$ and $\mathbf{a}_2(\mathbf{x}_{ik}, \mathbf{d}_{ik}, t)$ are bounded, and they satisfy

$$\begin{aligned} &\| \mathbf{a}_1(\mathbf{x}_{ik}, t) \|_\infty \leq \delta_1 \sqrt{\| \mathbf{s}_{ik} \|_2} \\ &\| \dot{\mathbf{a}}_2(\mathbf{x}_{ik}, \mathbf{d}_{ik}, t) \|_\infty \leq \delta_2 \end{aligned} \tag{3.15}$$

where δ_1 and δ_2 are positive but unknown.

Concerning Assumptions 1 and 2, the input-output dynamics of the sliding mode vector $\mathbf{s}_{ik}$ in (3.9) can have the form of

$$\dot{\mathbf{s}}_{ik} = \mathbf{a}(\mathbf{x}_{ik}, \mathbf{d}_{ik}, t) + \mathbf{b}_1(\mathbf{x}_{ik}, \mathbf{\Delta}_k, t)\varpi_k \tag{3.16}$$

Here $\mathbf{b}_1(\mathbf{x}_{ik}, \mathbf{\Delta}_k, t) = \mathbf{I}_2 + \Delta \mathbf{b}(\mathbf{x}_{ik}, \mathbf{\Delta}_k, t)\mathbf{b}_0^{-1}(\mathbf{x}_{ik}, t)$ is a 2×2 matrix and $\varpi_k = \mathbf{b}_0(\mathbf{x}_{ik}, t)\mathbf{u}_k$.

From Assumption 1, we have

$$1 - \gamma_1 \leq \| \mathbf{b}_1(\mathbf{x}_{ik}, \mathbf{\Delta}_k, t) \|_2 \leq 1 + \gamma_1 \tag{3.17}$$

Adaptive-Gain Super-Twisting Sliding Mode Design

According to the super-twisting law, the following sliding mode formation control approach is addressed. Considering the input-output dynamics (3.16), here the variable ϖ_k is related to the control input $\mathbf{u}_k$ so that their control design is equivalent to each other. Therefore, the following sliding mode control is formulated.

$$\varpi_k = \varpi_{k1} + \varpi_{k2} \tag{3.18}$$

Here

$$\begin{aligned} \varpi_{k1} &= -\alpha_k \sqrt{\| \mathbf{s}_{ik} \|_2}\, \mathrm{sgn}(\mathbf{s}_{ik}) \\ \dot{\varpi}_{k2} &= -\frac{\beta_k}{2} \mathrm{sgn}(\mathbf{s}_{ik}) \end{aligned} \tag{3.19}$$

where α_k and β_k are the adaptive gains to be deduced from the closed-loop system stability. The signum function $\mathrm{sgn}(\mathbf{s}_{ik})$ in (3.19) is determined by $\mathrm{sgn}(\mathbf{s}_{ik}) = \left[\mathrm{sgn}(s_{ik,1}) \quad \mathrm{sgn}(s_{ik,2}) \right]^T$.

According to (3.16) and (3.18), the input-output dynamics can be re-written as

$$\begin{aligned} \dot{\mathbf{s}}_{ik} &= -\alpha_k \sqrt{\| \mathbf{s}_{ik} \|_2}\, \mathbf{b}_1 \mathrm{sgn}(\mathbf{s}_{ik}) + \mathbf{a}_1 + \varpi_{k*} \\ \dot{\varpi}_{k*} &= -\frac{\beta_k}{2} \mathbf{b}_1 \mathrm{sgn}(\mathbf{s}_{ik}) + \dot{\mathbf{a}}_2 + \dot{\mathbf{b}}_1 \varpi_{k2} \\ \varpi_{k*}(0) &= 0 \end{aligned} \tag{3.20}$$

Here $\mathbf{a}_1(\mathbf{x}_{ik}, t)$, $\dot{\mathbf{a}}_2(\mathbf{x}_{ik}, \mathbf{d}_{ik}, t)$ and $\dot{\mathbf{b}}_1(\mathbf{x}_{ik}, \mathbf{\Delta}_k, t)$ are abbreviated by $\mathbf{a}_1$, $\dot{\mathbf{a}}_2$ and $\dot{\mathbf{b}}_1$ for brevity, and $\varpi_{k*} = \mathbf{a}_2 + \mathbf{b}_1 \varpi_{k2}$.

Assumption 3: $\dot{\mathbf{b}}_1 \varpi_{k2}$ in (3.20) is bounded but its boundary is unknown, that is,

$$\| \dot{\mathbf{b}}_1 \varpi_{k2} \|_\infty \leq \delta_3 \tag{3.21}$$

Here δ_3 is positive but unknown.

Substituting ϖ_{k2} in (3.18) into (3.21) yields

$$\|\dot{\mathbf{b}}_1\varpi_{k2}\|_\infty \leq \frac{\|\dot{\mathbf{b}}_1\|_\infty}{2}\int_0^t \beta_k \,\mathrm{d}t \leq \delta_3 \tag{3.22}$$

This fact indicates the adaptive gain β_k is bounded as well, i.e.,

$$|\beta_k| \leq \beta^* \tag{3.23}$$

Here β^* is positive but unknown.

Assumption 4: the adaptive gain α_k is bounded, that is,

$$|\alpha_k| \leq \alpha^* \tag{3.24}$$

Here α^* is positive but unknown.

Note that the uncertainties and disturbances of the input-output dynamics (17) exist in the terms $\mathbf{a}_2$ and $\mathbf{b}_1$. From Assumptions 2 and 3, the lumped boundary of the uncertain terms in (3.16) can be deduced from (3.15) and (3.24).

$$\|\dot{\chi}\|_\infty = \|\dot{\mathbf{a}}_2 + \dot{\mathbf{b}}_1\varpi_{k2}\|_\infty \leq \|\dot{\mathbf{a}}_2\|_\infty + \|\dot{\mathbf{b}}_1\varpi_{k2}\|_\infty \leq \delta_2 + \delta_3 = \delta_4 \tag{3.25}$$

where $\dot{\chi} = \dot{\mathbf{a}}_2 + \dot{\mathbf{b}}_1\varpi_{k2}$.

Until now, the formation control design has been equivalent to the deduction of the adaptive-gain laws in (3.18) that can make $\mathbf{s}_{ik}$ and $\dot{\mathbf{s}}_{ik}$ convergent to zero in finite time despite the uncertainties and disturbances, where the input-output dynamics are decided by (3.20) and the unknown boundaries δ_1, γ_1 and δ_4 are given by (3.13), (3.15) and (3.25). Finally, the adaptive gains are designed by

$$\dot{\alpha}_k = \begin{cases} \xi\sqrt{\frac{\gamma_1}{2}}\,\mathrm{sgn}(\|\mathbf{s}_{ik}\|_2 - \mu) & \text{if}\, \alpha_k > \alpha_m \\ \eta & \text{if}\, \alpha_k < \alpha_m \end{cases}$$
$$\beta_k = 2\varepsilon\alpha_k \tag{3.26}$$

Here γ_1 is determined by (14); ε, ξ and η are arbitrary positive constants; $\alpha_m > 0$ is an arbitrary small constant and the initial condition of α_k at t=0 satisfies $\alpha_k(0) > \alpha_m$.

Stability Analysis of the Closed-Loop Control System

Theorem 1: Consider the dynamics (3.20), where $\mathbf{a}_1$, $\mathbf{a}_2$ and $\mathbf{b}_1$ satisfy Assumptions 1 and 2 with the unknown gains δ_1, δ_2, and γ_1. The adaptive gains satisfy Assumptions 3 and 4 and they are determined by (3.26). Then, for any $\mathbf{x}_{ik}$ at t=0, there exist

- a parameter $\mu > 0$ so that α_k satisfies (3.27) if $\|\mathbf{s}_{ik}\|_2 > \mu$ at t=0;

$$\alpha_k > \frac{\delta_1(\lambda + 4\varepsilon^2) - \varepsilon(4\delta_4 + 1)}{\lambda(1-\gamma_1)} + \frac{[2\varepsilon\delta_1 - 2\delta_4 - \lambda - 4\varepsilon^2]^2}{12\varepsilon\lambda(1-\gamma_1)} \tag{3.27}$$

  Here λ is an arbitrary positive constant and δ_4 is determined by (3.25).
- a finite time $t_F > 0$ so that the sliding modes of $\mathbf{s}_{ik}$ are reached in the finite time t_F regarding to the adaptive-gain super-twisting sliding mode control method, that is, $\forall t > t_F$, $\exists$ $\|\mathbf{s}_{ik}\|_2 \le \eta_1$ and $|\dot{\mathbf{s}}_{ik}\|_2 \le \eta_2$. Here $\eta_1 > \mu$ and $\eta_2 > 0$.
- both α_k and β_k are bounded.

Proof:

Preparation

Define a 4×1 augmented vector $\mathbf{Z}$

$$\mathbf{Z} = \begin{bmatrix} \mathbf{Z}_1^T & \mathbf{Z}_2^T \end{bmatrix}^T = \begin{bmatrix} \sqrt{\| \mathbf{s}_{ik} \|_2} [\mathrm{sgn}(\mathbf{s}_{ik})]^T & [\varpi_{k*}]^T \end{bmatrix}^T \tag{3.28}$$

Here $\| \mathbf{Z}_1 \|_\infty = \sqrt{\| \mathbf{s}_{ik} \|_2}$, $\mathrm{sgn}(\mathbf{Z}_1) = \mathrm{sgn}(\mathbf{s}_{ik})$ and $\mathbf{Z}_2 = \varpi_{k*}$. Further, the time derivative of $\mathbf{Z}_1$ and $\mathbf{Z}_2$ can be written as

$$\begin{aligned} \dot{\mathbf{Z}}_1 &= \frac{1}{2} \frac{1}{\| \mathbf{Z}_1 \|_\infty} \left[-\alpha_k \mathbf{b}_1 \mathbf{Z}_1 + \mathbf{Z}_2 + \mathbf{a}_1 \right] \\ \dot{\mathbf{Z}}_2 &= -\frac{\beta_k}{2} \mathbf{b}_1 \frac{\mathbf{Z}_1}{\| \mathbf{Z}_1 \|_\infty} + \dot{\chi} \end{aligned} \tag{3.29}$$

From (3.29), we have

$$\begin{bmatrix} \dot{\mathbf{Z}}_1 \\ \dot{\mathbf{Z}}_2 \end{bmatrix} = \frac{1}{2 \| \mathbf{Z}_1 \|_\infty} \begin{bmatrix} -\alpha_k \mathbf{b}_1 & \mathbf{I}_2 \\ -\beta_k \mathbf{b}_1 & \mathbf{O}_2 \end{bmatrix} \begin{bmatrix} \mathbf{Z}_1 \\ \mathbf{Z}_2 \end{bmatrix} + \frac{1}{2 \| \mathbf{Z}_1 \|_\infty} \begin{bmatrix} \mathbf{a}_1 \\ 2 \| \mathbf{Z}_1 \|_\infty \dot{\chi} \end{bmatrix} \tag{3.30}$$

here $\mathbf{O}_2$ is a 2×2 zero matrix.

Further, considering (3.15) and (3.25), two bounded scalar functions $\rho_1(\mathbf{x}_{ik}, t)$ and $\rho_2(\mathbf{x}_{ik}, t)$ can be constructed as

$$\begin{aligned} \mathbf{a}_1 &= \rho_1(\mathbf{x}_{ik}, t) \sqrt{\| \mathbf{s}_{ik} \|_2} \, \mathrm{sgn}(\mathbf{s}_{ik}) = \rho_1(\mathbf{x}_{ik}, t) \mathbf{Z}_1 \\ \dot{\chi} &= \frac{\rho_2(\mathbf{x}_{ik}, t)}{2} \frac{\sqrt{\| \mathbf{s}_{ik} \|_2}}{\sqrt{\| \mathbf{s}_{ik} \|_2}} \mathrm{sgn}(\mathbf{s}_{ik}) = \frac{\rho_2(\mathbf{x}_{ik}, t)}{2} \frac{\mathbf{Z}_1}{\| \mathbf{Z}_1 \|_\infty} \end{aligned} \tag{3.31}$$

Then, we can have

$$\begin{aligned} 0 &< \rho_1(\mathbf{x}_{ik}, t) < \delta_1 \\ 0 &< \rho_2(\mathbf{x}_{ik}, t) < 2\delta_4 \end{aligned} \tag{3.32}$$

Substituting (3.31) into (3.30) yields

$$\begin{bmatrix} \dot{\mathbf{Z}}_1 \\ \dot{\mathbf{Z}}_2 \end{bmatrix} = \bar{\mathbf{A}} \begin{bmatrix} \mathbf{Z}_1 \\ \mathbf{Z}_2 \end{bmatrix} \tag{3.33}$$

Here $\bar{\mathbf{A}} = \frac{1}{2\|\mathbf{Z}_1\|_\infty} \begin{bmatrix} -\alpha_k \mathbf{b}_1 + \rho_1(\mathbf{x}_{ik},t)\mathbf{I}_2 & \mathbf{I}_2 \\ -\beta_k \mathbf{b}_1 + \rho_2(\mathbf{x}_{ik},t)\mathbf{I}_2 & \mathbf{O}_2 \end{bmatrix}$.

Consider a Lyapunov candidate as

$$V_0 = \mathbf{Z}^T \mathbf{P} \mathbf{Z} \tag{3.34}$$

Here $\mathbf{P} = \begin{bmatrix} (\lambda + 4\varepsilon^2)\mathbf{I}_2 & -2\varepsilon \mathbf{I}_2 \\ -2\varepsilon \mathbf{I}_2 & \mathbf{I}_2 \end{bmatrix}$ is a 4×4 positive definite matrix, $\lambda > 0$ and $\varepsilon > 0$.

From (3.33), the time derivative of V_0 in (3.34) can be written by

$$\dot{V}_0 = \dot{\mathbf{Z}}^T \mathbf{P} \mathbf{Z} + \mathbf{Z}^T \mathbf{P} \dot{\mathbf{Z}} = \mathbf{Z}^T \left(\bar{\mathbf{A}}^T \mathbf{P} + \mathbf{P} \bar{\mathbf{A}}^T \right) \mathbf{Z} \le -\frac{1}{2\|\mathbf{Z}_1\|_\infty} \mathbf{Z}^T \bar{\mathbf{Q}} \mathbf{Z} \tag{3.35}$$

Here $\bar{\mathbf{Q}} = \begin{bmatrix} \bar{\mathbf{Q}}_{11} & \bar{\mathbf{Q}}_{12} \\ \bar{\mathbf{Q}}_{21} & 4\varepsilon \end{bmatrix}$ is a 4×4 symmetric matrix, $\bar{\mathbf{Q}}_{11}$, $\bar{\mathbf{Q}}_{12}$ and $\bar{\mathbf{Q}}_{21}$ are given by

$$\bar{\mathbf{Q}}_{11} = 2\lambda\alpha_k \mathbf{b}_1 + 4\varepsilon(2\varepsilon\alpha_k - \beta_k)\mathbf{b}_1 - [2(\lambda + 4\varepsilon^2)\rho_1(\mathbf{x}_{ik},t) - 4\varepsilon\rho_2(\mathbf{x}_{ik},t)]\mathbf{I}_2 \tag{3.36}$$

$$\bar{\mathbf{Q}}_{12} = \bar{\mathbf{Q}}_{21} = (\beta_k - 2\varepsilon\alpha_k)\mathbf{b}_1 + [2\varepsilon\rho_1(\mathbf{x}_{ik},t) - \rho_2(\mathbf{x}_{ik},t)]\mathbf{I}_2 - (\lambda + 4\varepsilon^2)\mathbf{I}_2 \tag{3.37}$$

From the aspect of the stability of V_0, $\bar{\mathbf{Q}}$ is not only symmetric but also positive definite in the sense of Lyapunov. Therefore, we select

$\beta_k = 2\varepsilon\alpha_k$ in $\bar{\mathbf{Q}}_{11}$. Then, $\bar{\mathbf{Q}}$ can be positive definite and its minimal eigenvalue is $\lambda_{\min}(\bar{\mathbf{Q}}) \geq 2\varepsilon$ if α_k satisfies (3.27).

From (3.34), we have

$$\lambda_{\min}(\mathbf{P}) \| \mathbf{Z} \|_2^2 \leq \mathbf{Z}^T \mathbf{P} \mathbf{Z} = V_0 \leq \lambda_{\max}(\mathbf{P}) \| \mathbf{Z} \|_2^2 \tag{3.38}$$

Further, (3.39) can be deduced from (3.35) if (3.27) holds true.

$$\dot{V}_0 \leq -\frac{1}{2\| \mathbf{Z}_1 \|_\infty} \mathbf{Z}^T \bar{\mathbf{Q}} \mathbf{Z} \leq -\frac{2\varepsilon}{2\| \mathbf{Z}_1 \|_\infty} \mathbf{Z}^T \mathbf{Z} = -\frac{\varepsilon \| \mathbf{Z} \|_2^2}{\| \mathbf{Z}_1 \|_\infty} \tag{3.39}$$

In (3.28), the following inequality exists.

$$\| \mathbf{Z}_1 \|_\infty = \sqrt{\| \mathbf{s}_{ik} \|_2} \leq \| \mathbf{Z}_1 \|_2 \leq \| \mathbf{Z} \|_2 \leq \left(\frac{V_0}{\lambda_{\min}(\mathbf{P})} \right)^{1/2} \tag{3.40}$$

With regard to (3.38) and (3.40), (3.39) can be re-written by

$$\dot{V}_0 \leq -\frac{\varepsilon \lambda_{\min}^{1/2}(\mathbf{P})}{\lambda_{\max}(\mathbf{P})} V_0^{1/2} \tag{3.41}$$

Analysis

Now, the closed-looped control system stability will be presented in the sense of Lyapunov. Define a Lyapunov candidate as

$$V = V_0 + \frac{1}{2\gamma_1}(\alpha_k - \alpha^*)^2 + \frac{1}{2\gamma_2}(\beta_k - \beta^*)^2 \tag{3.42}$$

Here α^* and β^* are given in Assumptions 3 and 4. Concerning (3.35) and (3.41), the time derivative of V can have the form of

$$\dot{V} = \dot{V}_0 + \frac{1}{\gamma_1}(\alpha_k - \alpha^*)\dot{\alpha} + \frac{1}{\gamma_2}(\beta_k - \beta^*)\dot{\beta} \tag{3.43}$$

From (3.41), (3.43) can be written as

$$\begin{aligned}\dot{V} &\leq -\frac{\varepsilon\lambda_{\min}^{1/2}(\mathbf{P})}{\lambda_{\max}(\mathbf{P})}V_0^{1/2} + \frac{1}{\gamma_1}\varepsilon_\alpha\dot{\alpha}_k + \frac{1}{\gamma_2}\varepsilon_\beta\dot{\beta}_k \\ &= -\frac{\varepsilon\lambda_{\min}^{1/2}(\mathbf{P})}{\lambda_{\max}(\mathbf{P})}V_0^{1/2} - \frac{\xi_1}{\sqrt{2\gamma_1}}|\varepsilon_\alpha| - \frac{\xi_2}{\sqrt{2\gamma_2}}|\varepsilon_\beta| + \frac{1}{\gamma_1}\varepsilon_a\dot{\alpha}_k \\ &+ \frac{1}{\gamma_2}\varepsilon_\beta\dot{\beta}_k + \frac{\xi_1}{\sqrt{2\gamma_1}}|\varepsilon_\alpha| + \frac{\xi_2}{\sqrt{2\gamma_2}}|\varepsilon_\beta|\end{aligned} \tag{3.44}$$

Here $\varepsilon_\alpha = \alpha_k - \alpha^*$ and $\varepsilon_\beta = \beta_k - \beta^*$. It is apparent that both ε_α and ε_β are negtive or equal to zero according to Assumptions 3 and 4.

Since $(\tilde{a}^2 + \tilde{b}^2 + \tilde{c}^2)^{\frac{1}{2}} \leq |\tilde{a}| + |\tilde{b}| + |\tilde{c}|$, i it is concluded that

$$-\frac{\varepsilon\lambda_{\min}^{1/2}(\mathbf{P})}{\lambda_{\max}(\mathbf{P})}V_0^{1/2} - \frac{\xi_1}{\sqrt{2\gamma_1}}|\varepsilon_\alpha| - \frac{\xi_2}{\sqrt{2\gamma_2}}|\varepsilon_\beta| \leq -\eta_0 V^{1/2} \tag{3.45}$$

Here $\eta_0 = \min(\frac{\varepsilon\lambda_{\min}^{1/2}(\mathbf{P})}{\lambda_{\max}(\mathbf{P})}, \xi_1, \xi_2)$. Then, (3.44) can have the form of

$$\dot{V} \leq -\eta_0 V^{1/2} + \frac{1}{\gamma_1}\varepsilon_\alpha\dot{\alpha}_k + \frac{1}{\gamma_2}\varepsilon_\beta\dot{\beta}_k + \frac{\xi_1}{\sqrt{2\gamma_1}}|\varepsilon_\alpha| + \frac{\xi_2}{\sqrt{2\gamma_2}}|\varepsilon_\beta| \tag{3.46}$$

Since both ε_α and ε_β in (3.46) are equal to or less than zero, we can obtain

$$\dot{V} \leq -\eta_0 V^{1/2} - |\varepsilon_\alpha|(\frac{1}{\gamma_1}\dot{\alpha}_k - \frac{\xi_1}{\sqrt{2\gamma_1}}) - |\varepsilon_\beta|(\frac{1}{\gamma_2}\dot{\beta}_k - \frac{\xi_2}{\sqrt{2\gamma_2}}) \tag{3.47}$$

The motivation of designing the adaptive gains is to investigate a domain. The domain acts as a flag. The gains α_k and β_k can start dynamically reducing when the system trajectories come to the domain in finite time. Once the trajectories leave the domain, the gains start dynamically increasing in order to draw the trajectories back. Inspired by the methodology of sliding mode, we pick up the domain $\|\mathbf{s}_{ik}\|_2 \leq \mu$ as this flag. Thereafter, (3.47) is going to be deduced by different cases in accordance with such a flag.

Case 1: $\|\mathbf{s}_{ik}\|_2 > \mu$ and $\alpha_k > \alpha_m$ for all $t \geq 0$. Take (3.26) into account. With regard to this case, (3.48) can be deduced from (3.26).

$$\dot{\alpha}_k = \xi_1 \sqrt{\frac{\gamma_1}{2}} \tag{3.48}$$

Therefore, (3.47) becomes

$$\dot{V} \leq -\eta_0 V^{1/2} - |\varepsilon_\beta| \left(\frac{1}{\gamma_2}\dot{\beta}_k - \frac{\xi_2}{\sqrt{2\gamma_2}}\right) \tag{3.49}$$

Picking up $\varepsilon = \frac{\xi_2}{2\xi_1}\sqrt{\frac{\gamma_2}{\gamma_1}}$ in (3.13) and substituting it into the time derivative of $\beta_k = 2\varepsilon\alpha_k$ yields

$$\dot{\beta}_k = \xi_2 \sqrt{\frac{\gamma_2}{2}} \tag{3.50}$$

Finally, (3.49) becomes

$$\dot{V} \leq -\eta_0 V^{1/2} \tag{3.51}$$

Associated with the closed-loop system stability, (3.35) must be held true in order to have the positive definite matrix $\bar{\mathbf{Q}}$, meaning that α should be increased as its adaptive law (3.26) until (3.35) is satisfied in the finite time t_F. From this time t_F on, (3.51) will guarantee the convergence of this closed-loop system to the domain $\|\mathbf{s}_{ik}\|_2 \le \mu$.

Case 2: $\|\mathbf{s}_{ik}\|_2 \le \mu$. Take (3.26) into account. According to the motivation, α_k needs to be reduced in light of the adaptive law (3.26) so that it has a form of

$$\dot{\alpha}_k = \begin{cases} -\xi\sqrt{\frac{\gamma_1}{2}} & \text{if}\,\alpha_k > \alpha_m \\ \eta & \text{if}\,\alpha_k < \alpha_m \end{cases} \tag{3.52}$$

Meanwhile, picking up $\varepsilon = \frac{\xi_2}{2\xi_1}\sqrt{\frac{\gamma_2}{\gamma_1}}$ in (3.13) and substituting it into the time derivative of $\beta_k = 2\varepsilon\alpha_k$ yields

$$\dot{\beta}_k = \begin{cases} -\xi_2\sqrt{\frac{\gamma_2}{2}} & \text{if}\,\alpha_k > \alpha_m \\ \frac{\xi_2}{\xi_1}\sqrt{\frac{\gamma_2}{\gamma_1}}\eta & \text{if}\,\alpha_k < \alpha_m \end{cases} \tag{3.53}$$

Consequently, (3.47) becomes

$$\dot{V} \le \begin{cases} -\eta_0 V^{1/2} + 2|\varepsilon_\alpha|\frac{\xi_1}{\sqrt{2\gamma_1}} + 2|\varepsilon_\beta|\frac{\xi_2}{\sqrt{2\gamma_2}} & \text{if}\alpha_k > \alpha_m \\ -\eta_0 V^{1/2} - |\varepsilon_\alpha|(\frac{\eta}{\gamma_1} - \frac{\xi_1}{\sqrt{2\gamma_1}}) - |\varepsilon_\beta|(\frac{\xi_2}{\xi_1}\frac{\eta}{\sqrt{\gamma_1\gamma_2}} - \frac{\xi_2}{\sqrt{2\gamma_2}}) & \text{if}\alpha_k < \alpha_m \end{cases} \tag{3.54}$$

(3.54) indicates that the sign of the time derivative of V is indefinite so that $\|\mathbf{s}_{ik}\|_2$ may become larger than μ with the decrease of α_k and β_k.

Once $\|\mathbf{s}_{ik}\|_2$ becomes greater than μ. The condition defined in Case 1 will be immediately triggered. In (3.51), the time derivative of V is negative so that the closed-looped control system possesses the inherent stability and $\mathbf{s}_{ik}$ will enter the domain $\|\mathbf{s}_{ik}\|_2 < \mu$ again in finite time. This process continues back and forth till the control system become convergent. In this process, $\mathbf{s}_{ik}$ may deviate from the domain for a finite time but there always exists another domain in the real sliding modes of $\mathbf{s}_{ik}$.

$$\|\mathbf{s}_{ik}\|_2 \le \eta_1 \quad (\eta_1 > \mu) \tag{3.55}$$

Inside the domain $\|\mathbf{s}_{ik}\|_2 < \mu$, the value of $\|\dot{\mathbf{s}}_{ik}\|_2$ can estimated from (3.20) and (3.26).

$$\|\dot{\mathbf{s}}_{ik}\|_2 \le \left[(1-\gamma_1)\alpha_k(t_1) + \delta_1\right]\mu^{1/2} + \left[\varepsilon(1-\gamma_1)\alpha_k(t_1) + \delta_4\right](t_2 - t_1) = \bar{\eta}_2 \tag{3.56}$$

Here t_1 is the time instant when $\mathbf{s}_{ik}$ enters the domain $\|\mathbf{s}_{ik}\|_2 \le \mu$ and t_2 is the moment when $\mathbf{s}_{ik}$ leaves the domain. Once $\|\mathbf{s}_{ik}\|_2$ becomes $\mu \lhd \|\mathbf{s}_{ik}\|_2 < \eta_1$, we have

$$\|\dot{\mathbf{s}}_{ik}\|_2 \le (1+\gamma_1)\left(\eta_1^{1/2} + \varepsilon\right)\left(\alpha_k(t_2) + \xi_1\sqrt{\frac{\eta_1\gamma_1}{2}} + \varepsilon\right)(t_3 - t_2) + \delta_1\eta_1^{1/2} + \delta_4(t_3 - t_2) = \tilde{\eta}_2 \tag{3.57}$$

Here t_2 is the time instant when $\mathbf{s}_{ik}$ leaves the domain $\|\mathbf{s}_{ik}\|_2 < \mu$ and t_3 is the moment when $\mathbf{s}_{ik}$ enter the domain $\|\mathbf{s}_{ik}\|_2 < \mu$. Subsequently, (3.58) can be drawn from (3.56) and (3.57).

$$\|\dot{\mathbf{s}}_{ik}\|_2 \le \max\left(\tilde{\eta}_2,\ \bar{\eta}_2\right) = \eta_2 \tag{3.58}$$

From (3.55) and (3.57), there exist the real sliding modes, described by

$$\Omega=\{\mathbf{s}_{ik},\dot{\mathbf{s}}_{ik}:\ \|\mathbf{s}_{ik}\|_2\le\eta_1\quad\|\dot{\mathbf{s}}_{ik}\|_2\le\eta_2\qquad\eta_1>\mu\}\tag{3.59}$$

The existences of the sliding modes in (3.58) can be presented in theory but η_1 and η_2 cannot be gotten in advance before a real control process is carried out.

Further, a solution of (3.25) in the domain $\mu<\|\mathbf{s}_{ik}\|_2\le\eta_1$ can be gotten as

$$\alpha_k=\alpha_k(0)+\xi_1 t\sqrt{\frac{\gamma_1}{2}}\qquad 0\le t\le t_F\tag{3.60}$$

(3.60) indicates that the adaptive gain α_k is bounded so that the gain β_k is bounded on account of $\beta_k=2\varepsilon\alpha_k$. Inside the domain $\|\mathbf{s}_{ik}\|_2\le\mu$, the adaptive gains α_k and β_k can be decreased as the presented control design. Consequently, α_k and β_k are bounded.

Hitherto, the three pieces of results in Theorem 1 have been proven. In the next section, this presented control method will be carried out for the formation maneuvers of an uncertain multi-robot system made of such several leader-follower pairs.

IMPLEMENTATION

Multi-Robot Simulation Platform

In order to demonstrate the feasibility of the designed formation control method, a multi-robot system platform is established in this section. The platform contains three identical robots shown in Chapter 1, where the robot numbered by 1 is assigned as the leader and other two robots numbered by 2 and 3 become its followers. The number of this plat form is 3 so that the formation system is a typical small-scale one, indicating that some assumptions such as no collisions and no communication delay can easily be held true. The robot radius is set by r=0.05m. Considering this platform, the parameters Δ_1, Δ_2 and Δ_3 of the three robots has the form of

$$\Delta_1 = \Delta_2 = \Delta_3 = \begin{bmatrix} \bar{\Delta} & 0 \\ 0 & \bar{\Delta} \end{bmatrix} \tag{3.61}$$

Here $\bar{\Delta}$ is set by 0.3×rad−0.2 and rad is random between 0 and 1. Other parameters indicate the uncertainties and disturbances are formulated by

$$\begin{aligned} \pi_{1x} = \pi_{1y} = \pi_{1\theta} = 0.5\sin(2\pi t) \\ \pi_{2x} = \pi_{2y} = \pi_{2\theta} = 0.3\sin(2\pi t) \\ \pi_{3x} = \pi_{3y} = \pi_{3\theta} = 0.3\sin(2\pi t) \end{aligned} \tag{3.62}$$

Concerning each follower, the parameters of their formation controller are set by $\varepsilon = 1$, $\gamma_1 = 2$, $\xi = 2$, $\mu = 0.7$, $\alpha_m = 0.01$ and $\eta = \alpha_m$; their parameters of the sliding modes are set by $\mathbf{C}_1 = \begin{bmatrix} 400 & 0 \\ 0 & 400 \end{bmatrix}$ and $\mathbf{C}_2 = \begin{bmatrix} 56 & 0 \\ 0 & 56 \end{bmatrix}$. Considering the adaptive gains of the two followers, the initial values of α_2 and α_3 is picked up as 19 and 21, respectively. Meanwhile, the initial value of the super-twisting law in (18) is given by $\varpi_{22} = \varpi_{32} = 2$.

Simulation Results

In Figure 3.1, the multi-robot platform carries out the task of string formation when moving along a circular trajectory, where the red means the leader robot and the green and blue delegate the two followers. The initial postures of the three robots are allocated at

$$
\begin{aligned}
\mathbf{q}_1 &= [0.5\text{m} \quad 0\text{m} \quad 0.5\pi\text{rad}]^T \\
\mathbf{q}_2 &= [0.8\text{m} \quad -0.4\text{m} \quad 0\text{rad}]^T \\
\mathbf{q}_3 &= [1\text{m} \quad -0.5\text{m} \quad \pi\text{rad}]^T
\end{aligned} \tag{3.63}
$$

According to the initial postures and the formation task, the initial states of the formation dynamics (3.1) can be calculated as

$$
\begin{aligned}
\mathbf{x}_{12}(0) &= [0.525\text{m} \quad 0\text{m/s} \quad 0.8\pi\text{rad} \quad 0\text{rad/s}]^T \\
\mathbf{x}_{13}(0) &= [0.707\text{m} \quad 0\text{m/s} \quad 0.75\pi\text{rad} \quad 0\text{rad/s}]^T
\end{aligned} \tag{3.64}
$$

Similarly, the desired states of the two followers are assigned by

$$
\begin{aligned}
\mathbf{x}_{12}^d &= [0.13\text{m} \quad 0\text{m/s} \quad 0.5\pi\text{rad} \quad 0\text{rad/s}]^T \\
\mathbf{x}_{13}^d &= [0.26\text{m} \quad 0\text{m/s} \quad 0.5\pi\text{rad} \quad 0\text{rad/s}]^T
\end{aligned} \tag{3.65}
$$

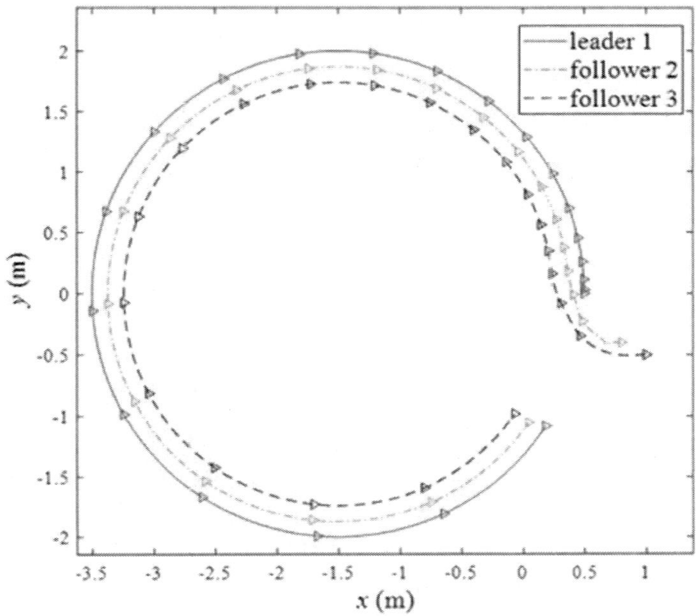

Figure 3.1. String formation of the platform when moving along a circular trajectory.

The leader's linear speed and its angular velocity are set as $v_1 = 0.5\text{m/s}$ and $\omega_1 = 1\text{rad/s}$.

Figure 3.2 demonstrates the state variables in (3.1) when the multi-robot system fulfills the formation task in Figure 3.1. From Figure 3.2, the presented control method can apparently improve the performance of the system state variables in (3.1).

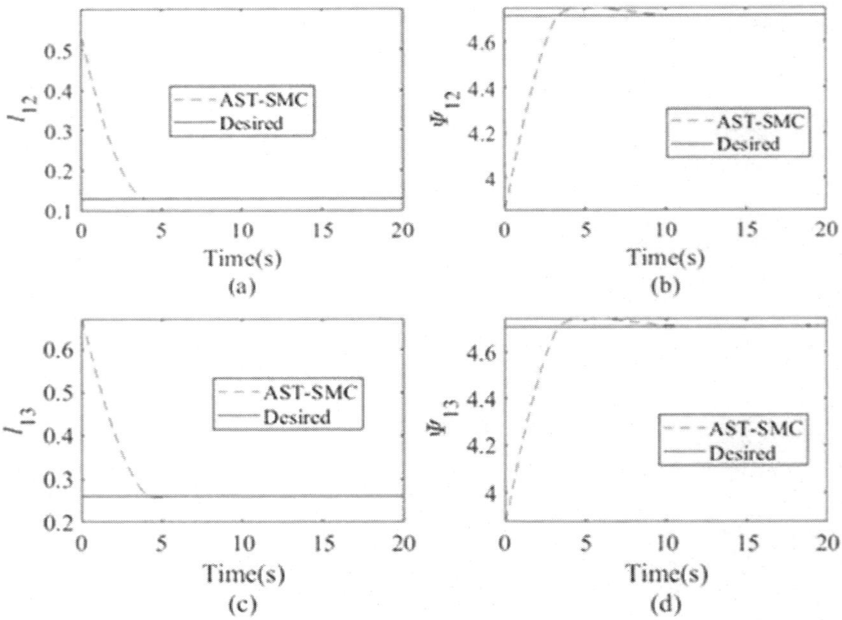

Figure 3.2. State variables by the designed control method. (a). l_{12} of follower 2, (b). ψ_{12} of follower 2, (c). l_{13} of follower 3 and (d). ψ_{13} of follower 3.

Shown in Figures 3.3 and 3.4, the adaptive-gain super-twisting sliding mode control method can decrease the chattering phenomenon effectively. In theory, the adaptive-gain super-twisting sliding mode control can completely compensate the disturbances and uncertainties entering the formation control system by the control channel. However, the formation dynamics (3.1) contains some disturbances and uncertainties that enter the control system by other channels. Therefore, the control inputs have to frequently switch to resist their adverse effects.

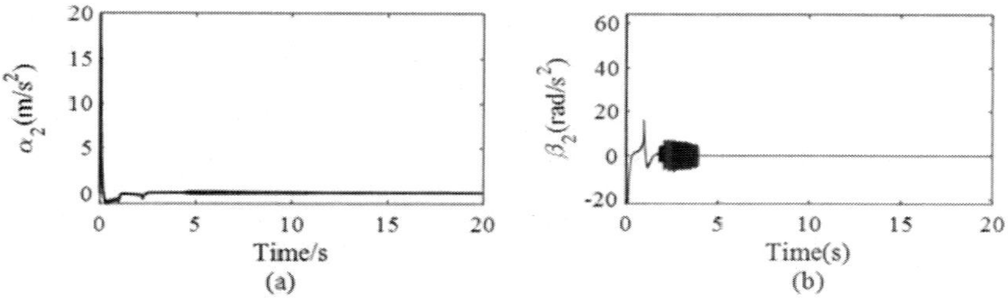

Figure 3.3. Accelerations and angular accelerations of follower 2 by the three control methods. (a). α_2 by the presented method and (b). β_2 by the presented method.

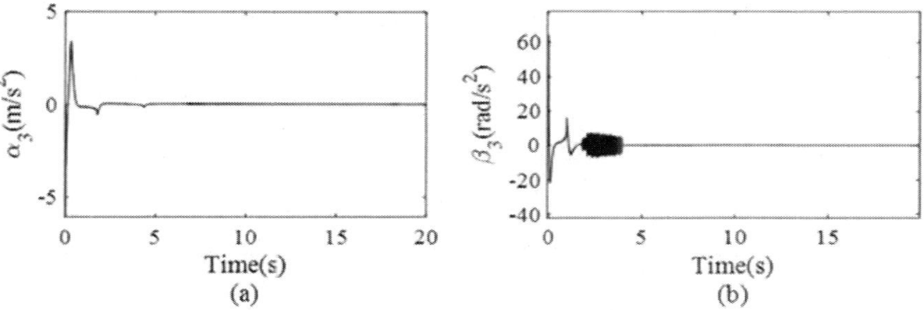

Figure 3.4. Accelerations and angular accelerations of follower 3 by the three control methods. (a). α_3 by the presented method and (b). β_3 by the presented method.

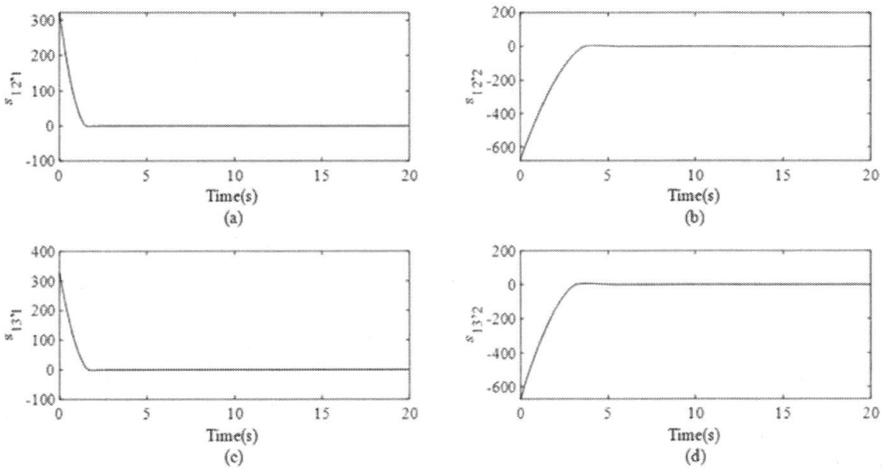

Figure 3.5. Sliding surfaces of the two followers.

Figure 3.5 illustrates the sliding surfaces of the adaptive-gain super-twisting sliding mode control method. The gains governed by the designed adaptive law (3.26) are demonstrated in Figure 3.6. As proven in Theorem 1, the sliding surfaces are convergent in infinite time although the time instants cannot be known in advance. Further, the curves of the gains in Figure 3.6 are not convergent. In fact, they are bounded as proven in Theorem 1.

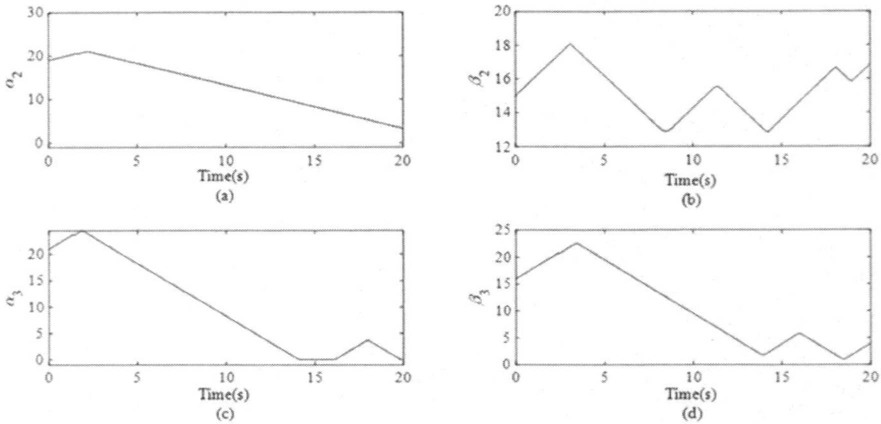

Figure 3.6. Adaptive gains of the two followers.

CONCLUSION

This chapter has concentrated on the formation control of multi-robot systems. In order to accomplish the formation task despite the inevitable disturbances and uncertainties, the super-twisting sliding mode control method is adopted. For the sake of dealing with the overestimate of the control gains, the adaptive laws of the gains are deduced. Theoretically, this adaptive-gain super-twisting sliding mode control method for the formation maneuvers is investigated in the sense of Lyapunov. Such a control method can guarantee the convergence of the sliding surfaces and make the adaptive gains bounded. Practically, the control method is applied to a multi-robot platform with three mobile robots. The numerical results illustrate the performance of the presented method has. The presented control method can be a solid support to solve the formation maneuvers of uncertain multi-robot systems.

REFERENCES

[1] Oh, K. K., Park, M. C., Ahn, H. S. (2015). A survey of multi-agent formation control. *Automatica*, 53, 424-440.

[2] Egerstedt, M., Hu, X. (2001). Formation constrained multi-agent control. *IEEE Transactions on robotics and automation*, 17(6), 947-951.

[3] Lin, Z., Wang, L., Han, Z., Fu, M. (2014). Distributed formation control of multi-agent systems using complex Laplacian. *IEEE Transactions on Automatic Control*, 59(7), 1765-1777.

[4] Mastellone, S., Stipanović, D. M., Graunke, C. R., Intlekofer, K. A., Spong, M. W. (2008). Formation control and collision avoidance for multi-agent non-holonomic systems: Theory and experiments. *The International Journal of Robotics Research*, 27(1), 107-126.

[5] Dong, X., Hu, G. (2016). Time-varying formation control for general linear multi-agent systems with switching directed topologies. *Automatica*, 73, 47-55.

[6] De Queiroz, M., Cai, X., Feemster, M. (2019). *Formation Control of Multi-Agent Systems: A Graph Rigidity Approach*. John Wiley & Sons.

[7] Alonso-Mora, J., Montijano, E., Nägeli, T., Hilliges, O., Schwager, M., Rus, D. (2019). Distributed multi-robot formation control in dynamic environments. *Autonomous Robots*, 43(5), 1079-1100.

[8] López-González, A., Campaña, J. M., Martínez, E. H., Contro, P. P. (2020). Multi robot distance based formation using Parallel Genetic Algorithm. *Applied Soft Computing*, 86, 105929.

[9] Li, Z., Yuan, Y., Ke, F., He, W., Su, C. Y. (2019). Robust Vision-Based Tube Model Predictive Control of Multiple Mobile Robots for

Leader–Follower Formation. *IEEE Transactions on Industrial Electronics*, 67(4), 3096-3106.

[10] Renzaglia, A., Briñón-Arranz, L. (2019). Search and Localization of a Weak Source with a Multi-robot Formation. *Journal of Intelligent & Robotic Systems*, 1-12.

[11] Yu, J., Ji, J., Miao, Z., Zhou, J. (2019). Neural network-based region reaching formation control for multi-robot systems in obstacle environment. *Neurocomputing*, 333, 11-21.

[12] Pan, Z., Wang, D., Deng, H., Li, K. (2019). A virtual spring method for the multi-robot path planning and formation control. *International Journal of Control, Automation and Systems*, 17(5), 1272-1282.

[13] Zong, C., Ji, Z., Tian, L., Zhang, Y. (2019). Distributed Multi-Robot Formation Control Based on Bipartite Consensus with Time-Varying Delays. *IEEE Access*, 7, 144790-144798.

[14] Zhang, X., Peng, Z., Yang, S., Wen, G., Rahmani, A. (2019). Distributed fixed-time consensus-based formation tracking for multiple nonholonomic wheeled mobile robots under directed topology. *International Journal of Control*, 1-10.

[15] Yang, H. A., Bao, X., Zhang, S., Wang, X. (2019). A Multi-Robot Formation Platform based on an Indoor Global Positioning System. *Applied Sciences*, 9(6), 1165.

[16] He, Y., Zhu, L., Sun, G., & Dong, M. (2019). Study on formation control system for underwater spherical multi-robot. *Microsystem Technologies*, 25(4), 1455-1466.

[17] Bono, A., Fedele, G., Franzè, G. (2019, September). A distributed model predictive control strategy for vehicle teams in uncertain narrowed environments. In *Proceedings of 2019 24th IEEE International Conference on Emerging Technologies and Factory Automation* (pp. 927-932). IEEE.

[18] Zou, Y., Wen, C., Guan, M. (2019). Distributed adaptive control for distance-based formation and flocking control of multi-agent systems. *IET Control Theory & Applications*, 13(6), 878-885.

[19] Liao, T. L., Chan, W. S., Yan, J. J. (2019). Distributed adaptive dynamic surface formation control for uncertain multiple quadrotor systems with interval type-2 fuzzy neural networks. *Transactions of the Institute of Measurement and Control*, 41(7), 1861-1879.

[20] Zhao, W., Zhang, H. (2019). Distributed optimal coordination control for nonlinear multi-agent systems using event-triggered adaptive dynamic programming method. *ISA Transactions*, 91, 184-195.

[21] Ferrara, A., Incremona, G. P., Cucuzzella, M. (2019). *Advanced and Optimization Based Sliding Mode Control: Theory and Applications*. Society for Industrial and Applied Mathematics.

[22] Edwards, C., Spurgeon, S. (1998). *Sliding mode control: theory and applications*. Crc Press.

[23] Defoort, M., Floquet, T., Kokosy, A., Perruquetti, W. (2008). Sliding-mode formation control for cooperative autonomous mobile robots. *IEEE Transactions on Industrial Electronics*, 55(11), 3944-3953.

[24] Cui, R., Chen, L., Yang, C., Chen, M. (2017). Extended state observer-based integral sliding mode control for an underwater robot with unknown disturbances and uncertain nonlinearities. *IEEE Transactions on Industrial Electronics*, 64(8), 6785-6795.

[25] Qian, D., Tong, S., Guo, J., Lee, S. (2015). Leader-follower-based formation control of nonholonomic mobile robots with mismatched uncertainties via integral sliding mode. *Proceedings of the Institution of Mechanical Engineers, Part I: Journal of Systems and Control Engineering*, 229(6), 559-569.

[26] Qian, D., Xi, Y. (2018). Leader–Follower Formation Maneuvers for Multi-Robot Systems via Derivative and Integral Terminal Sliding Mode. *Applied Sciences*, 8(7), 1045.

[27] Hui, L., Li, J. (2009). Terminal sliding mode control for spacecraft formation flying. *IEEE Transactions on Aerospace and Electronic Systems*, 45(3), 835-846.

[28] Liu, N., Ling, R., Huang, Q., Zhu, Z. (2015). Second-order super-twisting sliding mode control for finite-time leader-follower

consensus with uncertain nonlinear multiagent systems. *Mathematical Problems in Engineering*, 2015.

[29] Pukdeboon, C. (2012). Second-order sliding mode controllers for spacecraft relative translation. *Applied Mathematical Sciences*, 6(100), 4965-4979.

[30] Wee, S. G., Dai, Y., Kang, T. H., Lee, S. G. (2019). Variable formation control of multiple robots via VRc and formation switching to accommodate large heading changes by leader robot. *Advances in Mechanical Engineering*, 11(6), 1687814019857339.

[31] Khaledyan, M., de Queiroz, M. (2019). A formation maneuvering controller for multiple non-holonomic robotic vehicles. *Robotica*, 37(1), 189-211.

[32] Hu, J., Sun, X., He, L. (2020). Formation Tracking for Nonlinear Multi-agent Systems with Input and Output Quantization via Adaptive Output Feedback Control. *Journal of Systems Science and Complexity*, 1-25.

[33] Wang, R. (2020). Adaptive output-feedback time-varying formation tracking control for multi-agent systems with switching directed networks. *Journal of the Franklin Institute*, 357(1), 551-568.

In: A Closer Look at Formation Control
Editor: Dianwei Qian
ISBN: 978-1-53618-177-7

Chapter 4

INTEGRAL SLIDING MODE-BASED FORMATION MANEUVERS OF MULTI-AGENT ROBOTS

Dianwei Qian* and Fengyang Han
School of Control and Computer Engineering,
North China Electric Power University, Beijing, China

ABSTRACT

This chapter investigates a coordinative scheme of integral sliding mode control for a group of nonholonomic mobile robots with mismatched uncertainties. Such a scheme is based on leader-follower approach to formation maneuvers for the robots. Since the models are subject to uncertainties, the mismatched uncertainties are challenging because it cannot be suppressed by the invariance of the integral sliding mode control. To suppress the mismatched uncertainties when forming up the robots, an integral sliding mode controller is designed. The system stability is also analyzed in the light of Lyapunov's theory and a sufficient condition is proven such that the formation maneuvers for the robots with mismatched uncertainties is of locally asymptotical stability

* Corresponding Author's Email: dianwei.qian@ncepu.edu.cn.

in the sliding mode stage. To validate the feasibility of the controller, simulation results illustrate it can not only suppress the mismatched uncertainties, but also form up and maintain the robots in any trajectory Some numerical results are also conducted by a test bed made of three nonholonomic mobile robots.

Keywords: leader-follower formation, multiple robots, uncertainties, integral sliding mode control

INTRODUCTION

Coordination among multiple mobile robots is an important and promising research field [1, 2]. The ability to coordinate such systems is of merit over a single robot system, including but not limited to efficiency, effectiveness and economy [3]. There are many cooperative tasks of multiple mobile robots like formation, coverage, and target aggregation. Formation control is inspired by biological phenomena and it is a branch among all the topics of study in this field [4, 5]. The problem of formation control means to control a group of robots to form up and move in specified geometrical shapes [6]. The control problem is attractive because it is rooted in real-world applications. Many cooperative tasks need the robots to maintain some prescribed formations when moving [7, 8].

Up to now, several cooperative mechanisms have been reported on the formation design of multiple mobile robots [9]. These mechanisms can be roughly categorized into behavior-based approach, virtual structure approach and leader-follower approach [10]. The idea behind the behavior-based approach is that the desired behaviors are assigned to each robot and that the final action of each robot is achieved by weighing the relative importance of each behavior [11]. Such an approach is subject to complex mathematical analyses so that the system performance and stability cannot be guaranteed. With the virtual structure approach, the formation maneuvers are treated as a single virtual rigid structure [12]. Instead of each single robot, the desired trajectories are assigned to the entire formation as a whole. The formation behavior in this case is exactly predictable but the approach may result in any single point of failure for the entire formation [13, 14]. In the leader-follower approach, one robot of the formation is designed as the leader and others are the followers. The leader robot moves along a predefined trajectory while the follower robots keep track of the leader robot and maintain the desired distance and the desired relative bearing angle. The main criticism to the approach is that it depends heavily on the leader for achieving the formation [15]. Nevertheless, such a hierarchical approach has blossomed because a reference trajectory is clearly defined by the leader and the internal formation stability is induced by the individual robots' control laws [16].

Based on the leader-follower cooperative mechanism, various control methods have been presented to form up and maintain a group of robots in predefined trajectories, i.e., predictive control [17], adaptive control [18], feedback linearization [19], back-stepping [20], sliding mode control [21], intelligent control [22], decentralized control [23], to name but a few. See Ahn's report for a complete review on recent philosophies about the leader-follower formation control for multiple mobile robots [24]. Among the diversity of control methods, sliding mode control (SMC) has been paid more and more attention [25]. SMC is a nonlinear feedback control method. The main strength of SMC is its robustness. On the sliding mode stage, the SMC control system is of invariance to matched uncertainties, which means the control system is not sensitive to uncertainties entering the system through the control channel. Additionally, the sliding mode stage is reached in finite time (better than an asymptotic behavior) [26].

So far, several control methods on basis of SMC have been investigated to solve the formation problem of a group of robots. Chang et al. in [27] presented a fuzzy SMC controller for the decentralized formation problem for multiple robots, where disturbances and uncertainties were discussed in consensus problems and the system stability was analyzed by the concepts of graph theory. Zhao and Zou [28] developed the terminal SMC principle for formation maneuvers of multiple mobile robots. To form up a group of robots, Fu, Jian and Yin [29] addressed a back-stepping SMC algorithm. Defoort et al. [30] designed first-order and second-order SMC controllers and illustrated the feasibility of their controllers by a test bed. Park et al. [31] constructed a dynamic robot model and employed the SMC law to achieve the formation control for mobile robots with uncertainties. In the presence of uncertainties, one common assumption in the mentioned literature is that they are matched. However, most physical systems do not satisfy the matched condition of SMC in practice, which means all the uncertainties cannot be suppressed by the invariance of SMC. It is necessary to consider the effects of mismatched uncertainties on the system stability and performance.

In contrast with the conventional SMC, the integral SMC can guarantee robustness in the whole state space because the system trajectory always starts from the sliding surface. Such a SMC method has been successfully applied to power systems [32], aerospace engineering [33], vehicle design [34], etc. This paper focuses on the development of an integral SMC method for the leader-follower formation control of multiple nonholonomic mobile robots. Mathematical models of the robots are subject to uncertainties. In the mentioned literature about formation maneuvers, uncertainties are not strictly categorized into matched uncertainties and mismatched ones. They are usually assumed as matched or bounded ones. To attack the effects of mismatched uncertainties, an integral SMC controller is developed to resist them. Stability analysis about the control system is considered. Simulation and experimental results illustrate the integral SMC control method can not only suppress the uncertainties, but also achieve the task of formation maneuvers.

The remainder of this paper is as follows. In Section 2, models of a single robot are depicted and the mechanism of the lead-follower approach is formulated. Section 3 presents the integral SMC method and the system stability analysis. Simulation results in Section 4 and experimental results in Section 5 illustrate the feasibility and validity of the proposed strategy, respectively. Finally, conclusions in section 6 are drawn.

MATHEMATICAL MODEL

Recall the dynamics (1.8). Define $\mathbf{x}=\left[x_1\ x_2\ x_3\ x_4\right]^T$, where $x_1=l_{ik}$, $x_2=\dot{l}_{ik}$, $x_3=\psi_{ik}$ and $x_4=\dot{\psi}_{ik}$. Then, the following state equations are deduced from (1.8).

$$\dot{\mathbf{x}}=\mathbf{A}\mathbf{x}+\mathbf{B}\mathbf{u}_k+\mathbf{d}(t) \tag{4.1}$$

where $\mathbf{A}=\begin{bmatrix}0&1&0&0\\0&0&0&0\\0&0&0&1\\0&0&0&0\end{bmatrix}$, $\mathbf{B}=\begin{bmatrix}0&0\\\cos\varphi_{ik}&r\sin\varphi_{ik}\\0&0\\-\dfrac{\sin\varphi_{ik}}{l_{ik}}&\dfrac{r\cos\varphi_{ik}}{l_{ik}}\end{bmatrix}$, $\mathbf{d}(t)=\mathbf{B}\Delta_k\mathbf{u}_k+\mathbf{L}(\mathbf{I}_2+\mathbf{\Delta}_i)\mathbf{u}_i+\mathbf{F}+\tilde{\mathbf{P}}$,

$\mathbf{L}=\begin{bmatrix}0&0\\-\cos\varphi_{ik}&0\\0&0\\\dfrac{\sin\varphi_{ik}}{l_{ik}}&-1\end{bmatrix}$, $\mathbf{F}=\left[0\quad F_1\quad 0\quad F_2\right]^T$ and $\mathbf{P}=\left[0\quad P_1\quad 0\quad P_2\right]^T$.

Note that the term $\mathbf{d}(t)$ depicting the uncertainties of the leader-follower cooperation mechanism consists of 4 parts, where $\mathbf{B}\Delta_k$, meaning the parameter variations of the follower robot k, is matched. Other 3 parts are not well-defined. In the mentioned literature, there are two methods to deal with them. One is to assume they are matched. But the assumption is really hard to ne satisfied because of their complex mathematical expressions. Thus, it is more general to assume the 3 parts of uncertainties are mismatched. The other is to assume they are bounded. The assumption is also slightly rough because it is hard to analyze the system stability in the sliding mode stage. This paper focuses on an integral SMC law for the robots' formation maneuvers with mismatched uncertainties. It is well-known that the invariance of the SMC methodology is able to resist the part of matched uncertainties. If the mismatched uncertainties are injected to the robots' formation, the system stability cannot be guaranteed by the integral SMC law. Thus, it is necessary to analyze the stability of the integral SMC control system.

DESIGN AND CLOSED-LOOP STABILITY

Control Design for the Followers

According to the methodology of SMC, the system state trajectory in the period of time before reaching the sliding surface is called the reaching stage and the system trajectory sliding along the sliding surface to the origin is the sliding mode stage. The invariance of the control system to matched uncertainties exists in the sliding mode but not during the reaching phase. The Integral SMC with an integral sliding surface is power to resist matched uncertainties. To develop such an integral SMC controller for the formation control of multiple robots, it is inevitable to define the integral sliding surface (4.2).

$$\mathbf{s} = \begin{bmatrix} s_1 \\ s_2 \end{bmatrix} = \mathbf{G}(\mathbf{x} - \mathbf{x}_0) - \int_0^t \mathbf{G}(\mathbf{A} - \mathbf{BK})x(\tau)\ \mathrm{d}\tau \tag{4.2}$$

here $\mathbf{x}_0$ is the initial state vector at $t = 0$, $\mathbf{s} \in \mathcal{R}^2$ is the vector of the integral sliding surface. $s_1 \in \mathcal{R}^1$ and $s_2 \in \mathcal{R}^1$ are scalars. $\mathbf{G}$ and $\mathbf{K}$ is 2×4 constant matrixes, where $\mathbf{G}$ is selected to guarantee $\mathbf{GB}$ is nonsingular and $\mathbf{K}$ is designed via pole assignment such that the eigenvalues of matrix $\mathbf{A} - \mathbf{BK}$ are less than zero.

Note that 1) $\mathbf{s}(0) = \mathbf{0}$ exists at $t = 0$, where $\mathbf{0} = [0\ 0]^T \in \mathcal{R}^2$, so that the sliding mode takes place at $t = 0$; 2) $\text{rank}\left[\mathbf{B}\ \mathbf{BA}\right] = 4$ in (4.1) so that the formation system is controllable and the system poles can be assigned.

Pointed out by V. I. Utkin, only the equivalent control law of the integral SMC law exists in the sliding mode stage and the equivalent control law ensures the system trajectory stays on the surface after reaching the sliding mode stage.

If $\mathbf{d}(t)$ is assumed to be matched, then $\mathbf{d}(t)=\mathbf{Bd}'(t)$ in (4.1) where $\mathbf{d}'(t)$ is bounded, i.e., a known constant d_0 exists so that $\|\mathbf{d}'(t)\|_\infty \le d_0$. According to the reachability condition of SMC, the integral SMC law for the formation system with matched uncertainties can be formulated. In the sliding mode stage, this control law can guarantee the control system is of asymptotical stability to resist matched uncertainties. The deduction is trivial and it is a standard process introduced by V. I. Utkin. However, the system stability in the sliding mode stage cannot be guaranteed by the standard integral SMC design any longer because the mismatched uncertainties are taken into account in (4.1). Thus, the system stability in the sliding mode stage should be analyzed at first.

Assumption 1: $\mathrm{rank}\left[\mathbf{B},\mathbf{d}(t)\right]\ne \mathrm{rank}\left[\mathbf{B}\right]$

Assumption 2: $\|\mathbf{d}(t)\|_\infty \le \bar{d}$ where $\bar{d}$ is a known positive constant.

Differentiating $\mathbf{s}(t)$ in (4.2) with respect to time t and substituting (4.1) into the derivative of (4.2) yield

$$\dot{\mathbf{s}}(t)=\mathbf{G}\dot{\mathbf{x}}-\mathbf{G}(\mathbf{A}-\mathbf{B})\mathbf{x}=\mathbf{GBKx}+\mathbf{GBu}_k+\mathbf{Gd}(t) \tag{4.3}$$

Under Assumption 1, the equivalent control in the sliding mode stage can be derived from $\dot{\mathbf{s}}=\mathbf{0}$ in (4.3).

$$\mathbf{u}_{keq}=-\mathbf{Kx}(t)-(\mathbf{GB})^{-1}\mathbf{Gd}(t) \tag{4.4}$$

here $\mathbf{u}_{keq}$ is the equivalent control law of the follower robot k. Substituting (4.2) into (4.4), the state equations with mismatched certainties in the sliding mode stage can be written as

$$\dot{\mathbf{x}}=(\mathbf{A}-\mathbf{BK})\mathbf{x}+[\mathbf{I}_4-\mathbf{B}(\mathbf{GB})^{-1}\mathbf{G}]\mathbf{d}(t) \tag{4.5}$$

here $\mathbf{I}_4$ is a 4×4 identity matrix. Define

$$\mathbf{A} = \mathbf{A} - \mathbf{BK} \tag{4.6}$$

$$\mathbf{D} = [\mathbf{I}_4 - \mathbf{B}(\mathbf{GB})^{-1}\mathbf{G}]\mathbf{d}(t) \tag{4.7}$$

Then (4.5) can be re-written as

$$\dot{\mathbf{x}} = \mathbf{Ax} + \mathbf{D} \tag{4.8}$$

Since a gain matrix $\mathbf{K}$ exists for stable $\mathbf{A}$, there is a symmetric positive definite matrix $\mathbf{M}$ for the following Lyapunov equation.

$$\mathbf{A}^T\mathbf{N} + \mathbf{NA} = -\mathbf{M} \tag{4.9}$$

where $\mathbf{N}$ is the solution of (4.9) for a given positive definite symmetric matrix $\mathbf{M}$.

Theorem 1: $\forall \mathbf{x} \in \mathcal{B}(\sigma)$, the dynamic system (4.2) in the sliding mode stage is stable at any time, where $\mathcal{B}(\sigma)$ is the complement of the closed ball centered at $\mathbf{x} = [0 \;\; 0 \;\; 0 \;\; 0]^T$ with its radius.

$$\sigma = \frac{2\bar{d}\|\mathbf{N}\|_\infty}{\lambda_{\min}(\mathbf{M})} \tag{4.10}$$

where $\lambda_{\min}(\mathbf{M})$ is the minimum eigenvalue of $\mathbf{M}$.

Proof: Take the positive definite function $V_1(t) = \mathbf{x}^T\mathbf{Nx}$ into account as a Lyapunov candidate function. Then differentiating $V_1(t)$ with respect to time t and substituting (4.8) into the derivative of $V_1(t)$ yield

$$\begin{aligned}\dot{V}_1(t) &= \mathbf{x}^T(\mathbf{A}^T\mathbf{N} + \mathbf{N}\mathbf{A})\mathbf{x} + \mathbf{D}^T\mathbf{N}\mathbf{x} + \mathbf{x}^T\mathbf{N}\mathbf{D} \\ &= -\mathbf{x}^T\mathbf{M}\mathbf{x} + \mathbf{D}^T\mathbf{N}\mathbf{x} + \mathbf{x}^T\mathbf{N}\mathbf{D}\end{aligned} \tag{4.11}$$

According to the Assumption 2, (4.12) exists.

$$\begin{aligned}&\|\mathbf{D}\|_\infty = \|[\mathbf{I}_4 - \mathbf{B}(\mathbf{GB})^{-1}\mathbf{G}]\mathbf{d}(t)\|_\infty \\ &\le \|[\mathbf{I}_4 - \mathbf{B}(\mathbf{GB})^{-1}\mathbf{G}]\|_\infty \| \mathbf{d}(t)\|_\infty \\ &\le \|[\mathbf{I}_4 - \mathbf{B}(\mathbf{GB})^{-1}\mathbf{G}]\|_\infty \bar{d}\end{aligned} \tag{4.12}$$

so that $\dot{V}_1(t)$ in (4.11) becomes

$$\dot{V}_1(t) \le -\lambda_{\min}(\mathbf{M})\| \mathbf{x}\|_\infty^2 + 2\bar{d}\| \mathbf{N}\|_\infty \| \mathbf{x}\|_\infty \tag{4.13}$$

The Lyapunov function $V_1(t)$ satisfies $\dot{V}_1(t) \le 0$ once the system states enter into the closed ball $\mathcal{B}(\sigma)$ and the eigenvalue $\lambda_{\min}(\mathbf{M}) > 0$. Thus, the leader robot k with mismatched uncertainties is stable in the sliding mode stage.

In contrast to an asymptotically stable control system for the robots with matched uncertainties, Theorem 1 reveals the control system is locally asymptotically stable for the robots with mismatched uncertainties. In some mentioned literature about the SMC-based formation control, they did not touch the topic of the system stability in the sliding mode stage under the effects of mismatched uncertainties and the uncertainties were only assumed to be bounded as Assumption 2. Theorem 1 explains the integral SMC formation control system is locally asymptotically stable under the effects of mismatched uncertainties. Compared with the matched uncertainties, the mismatched uncertainties dramatically affect and change the system stability in the sliding mode stage. The contents revealed by Theorem 1 have not bee paid enough attention theretofore. To the best of the authors' knowledge, it is the first time to obtain such a stability theorem about the effects of the mismatched uncertainties in the sliding mode stage for the SMC-based formation.

The system trajectory cannot keep on and slide along the sliding surface perfectly due to the effects of mismatched uncertainties although the sliding mode stage of the integral SMC law starts at $t=0$ as mentioned. To attack the issue, it is necessary to consider the reachability condition of the integral sliding surface. In [25], such a condition is described as

$$\frac{\mathbf{s}^T\dot{\mathbf{s}}}{\|\mathbf{s}\|_\infty}<0 \tag{4.14}$$

Theorem 2: Take Assumptions 1 and 2 into accounts. The reachability condition (4.14) is satisfied if the integral SMC control law is designed as

$$\mathbf{u}_k=-\mathbf{Kx}-(\mathbf{GB})^{-1}\|\mathbf{G}\|_\infty\bar{d}-(\mathbf{GB})^{-1}\kappa\mathrm{sgn}(\mathbf{s})-(\mathbf{GB})^{-1}\eta\mathbf{s} \tag{4.15}$$

where κ and η are positive constants and the signum function sgn($\mathbf{s}$) is defined as

$$\mathrm{sgn}(\mathbf{s})=[\mathrm{sgn}(s_1)\ \ \mathrm{sgn}(s_2)]^T \tag{4.16}$$

Proof: Consider the following Lyapunov function $V_2(t)=\|\mathbf{s}\|$. Differentiate V_2 with respect to time t. The derivative of V_2 is obtained as

$$\dot{V}_2(t)=\frac{\mathbf{s}^T\dot{\mathbf{s}}}{\|\mathbf{s}\|_\infty} \tag{4.17}$$

Note (4.17) is just the left term of the reachability condition inequality (4.14). Substitute (4.3) into (4.17). Then (4.18) is drawn.

$$\dot{V}_2(t)=\frac{\mathbf{s}^T}{\|\mathbf{s}\|_\infty}[\mathbf{GBKx}+\mathbf{GBu}_k+\mathbf{Gd}(t)] \tag{4.18}$$

Further, substituting (4.15) in (4.18) gives

$$\begin{aligned}\dot{V}_2(t) &= \frac{\mathbf{s}^T}{\|\mathbf{s}\|_\infty}[-\| G\|_\infty \bar{d} + \mathbf{Gd}(t) - \kappa \text{sgn}(\mathbf{s}) - \eta \mathbf{s}] \\ &\leq \frac{\mathbf{s}^T}{\|\mathbf{s}\|_\infty}\{-\| G\|_\infty [\bar{d} - \mathbf{d}(t)] - \kappa \text{sgn}(\mathbf{s}) - \eta \mathbf{s}\}\end{aligned} \tag{4.19}$$

According to Assumption 2, $\dot{V}_2(t) \leq -\kappa < 0$ exists, i.e., $\frac{\mathbf{s}^T \dot{\mathbf{s}}}{\|\mathbf{s}\|_\infty} < 0$.

Theorems 1 and 2 give theoretical insights into the designed integral SMC method for the formation problem of multiple mobile robots with mismatched uncertainties. Note the conclusion drawn from Theorem 2 is more general because it does not depend on the attributes of the uncertainties. The slight difference between matched uncertainties and mismatched ones is that (4.17) can be further simplified according to the matched condition if the uncertainties are matched. The integral SMC law of the follower robot k has been developed in (4.15). The control system stability of such a follower in the sliding mode stage has also been presented in Theorem 1. Note that $\mathbf{u}_i$ exists in (4.2). Although $\mathbf{u}_i$ is treated as disturbances for the follower k, it is the control input of the leader robot i. The control design of the leader robot i is considered in the following subsection.

Control Design for the Leader

As mentioned, the task of the leader robot i is to track a predefined trajectory. Such a control problem can be treated as trajectory tracking of a single robot. There has been a great deal of work in this field and different control methods have been developed. Biglarbegian [35] even directly assumed the leader robot is well-controlled by a developed technology. Here, we just design a kinematic-based sliding mode controller for trajectory tracking of the leader.

NUMERICAL RESULTS

Simulation Platform

To demonstrate the validity, feasibility and robustness of the designed integral SMC controller for the formation of nonholonomic mobile robots with mismatched uncertainties, simulations are performed and results are discussed in this section.

Consider a multiple robot system consisting of a leader and two follower robots. The three mobile robots are the same as each other. The leader robot is with the kinematic-based sliding mode controller [35] to track a predefined trajectory. The two follower robots each are with the integral SMC controller (4.15) to track the lead robot. The formation objective is to make the two followers maintain a desired distance and orientation with respect to the leader in presence of mismatched uncertainties.

In (4.1), both of the parameter variations matrixes are considered as

$$\Delta_i = \Delta_k = \begin{bmatrix} 0.3\text{rand}() - 0.15 & 0 \\ 0 & 0.3\text{rand}() - 0.15 \end{bmatrix} \tag{4.20}$$

here rand() is a Matlab command and it can generate a uniformly distributed random number. (4.20) mean the variations on the mass and the moment of inertia are $\pm 15\%$ of the nominal parameters in the dynamical process. The reason that we consider the 15% variation is that each robot is driven by two DC motors. The 15% parameter variation is enough to cover the regular parameter perturbation. Too many variations may result in the motor overload in practice.

Further, it is apparent that $\mathbf{P}$ in (4.1) is related to the term $\boldsymbol{\pi}_n$ and in (4.1). Note the terms $\pi_{ix} - \pi_{kx}$ and $\pi_{iy} - \pi_{ky}$ exist in $\mathbf{P}$, meaning that it is not representative to define $\boldsymbol{\pi}_i = \boldsymbol{\pi}_k$. To depict the term $\boldsymbol{\pi}_n$, different functions are assigned to the leader and the two followers, defined by

$$\pi_{ix} = \pi_{iy} = \pi_{i\theta} = 0.5\sin(2\pi t)$$
$$\pi_{kx} = \pi_{ky} = \pi_{k\theta} = 0.2\sin(\pi t) \quad (4.21)$$

It is apparent that the term $\mathbf{d}(t)$ depicting the uncertainties in (4.1) is mismatched by submitting (4.20) and (4.21) into (4.1).

As mentioned, the matrix $\mathbf{K}$ should be predefined so that the eigenvalues of matrix $\mathbf{A} - \mathbf{BK}$ in (4.3) are those in the specified vector $\lambda = [-1.750 \pm 0.661j \;\; -1.450 \pm 0.893j]^T$. Finally, the matrix $\mathbf{K}$ is formulated by

$$\begin{bmatrix} 3.5\cos\varphi_{ik} & 3.5\cos\varphi_{ik} & -2.9l_{ik}\sin\varphi_{ik} & -2.9l_{ik}\sin\varphi_{ik} \\ \frac{3.5}{r}\sin\varphi_{ik} & \frac{3.5}{r}\sin\varphi_{ik} & \frac{2.9l_{ik}}{r}\cos\varphi_{ik} & \frac{2.9l_{ik}}{r}\cos\varphi_{ik} \end{bmatrix} \quad (4.22)$$

Another sliding surface matrix $\mathbf{G}$ are selected by $\mathbf{G} = \mathbf{K}$. Other parameters of the integral SMC controller for the follower robot k are determined by $\kappa = 10.00$ and $\eta = 25.00$.

Circle Trajectory

Take a circle trajectory into account for the multiple robot system. The initial location vector for the leader is set by $\mathbf{q}_{i1} = [0.5\text{m} \;\; 0\text{m} \;\; \frac{\pi}{2}\text{rad}]^T$. The initial location vectors of the two followers are set by $\mathbf{q}_{k1} = [0.9\text{m} \;\; -0.3\text{m} \;\; 0\text{rad}]^T \mathbf{q}_{k2} = [1\text{m} \;\; -0.5\text{m} \;\; \pi\text{rad}]^T$, respectively.

According to the above three initial location vectors, the initial distance and orientation of the two followers with respect to the leader in (4.1) are calculated as $x_{k1} = [0.5\text{m} \;\; 0\text{m/s} \;\; 3.2\pi/4\text{rad} \;\; 0\text{rad/s}]^T$ and $x_{k2} = [0.707\text{m} \;\; 0\text{m/s} \;\; 3\pi/4\text{rad} \;\; 0\text{rad/s}]^T$. The desired system states in (4.2) are set by $x^d{}_{k1} = [0.13\text{m} \;\; 0\text{m/s} \;\; \pi\text{rad} \;\; 0\text{rad/s}]^T$ and $x^d{}_{k2} = [0.26\text{m} \;\; 0\text{m/s} \;\; \pi\text{rad} \;\; 0\text{rad/s}]^T$.

Consider the motor load, the constraint of the angular acceleration is $|\beta_i| \le 30\text{rad/s}^2$.

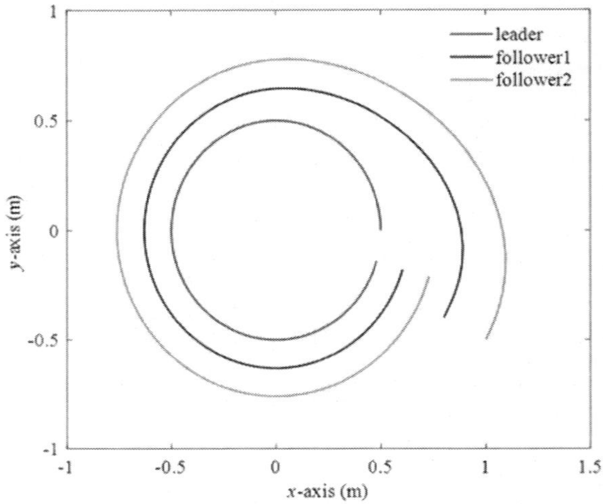

Figure 4.1. Straight line formation of the multiple robot system while moving in a circular trajectory.

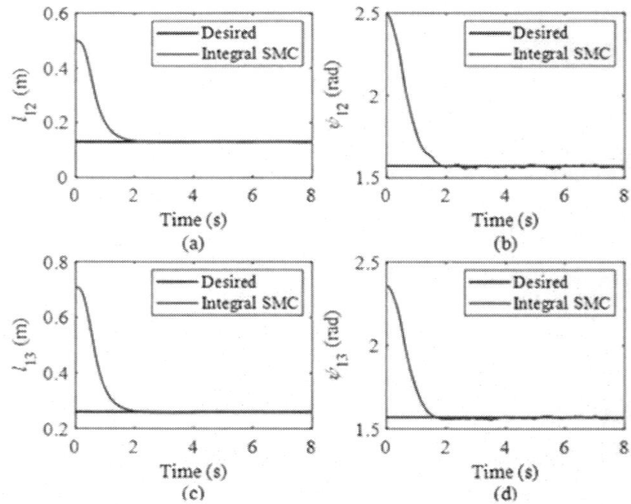

Figure 4.2. Distances and relative bearing angles of the two followers. (a). l_{12}, (b). ψ_{12}, (c). l_{13} and (d). ψ_{13}.

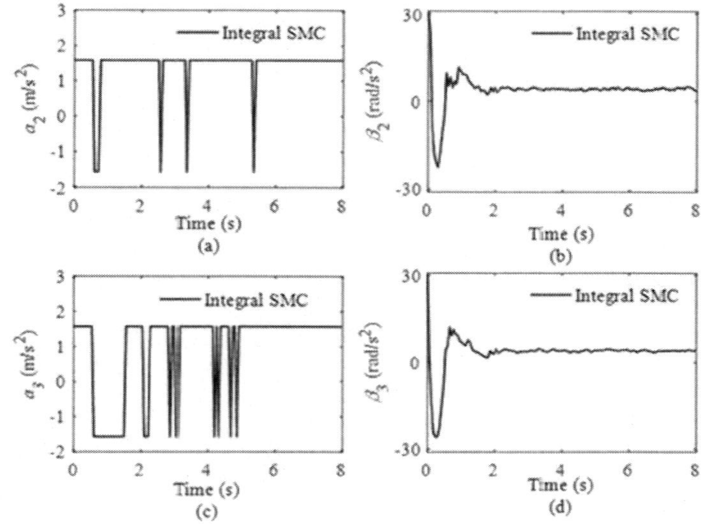

Figure 4.3. Control inputs of the two followers. (a). α_2, (b). β_2, (c). α_3 and (d). β_3.

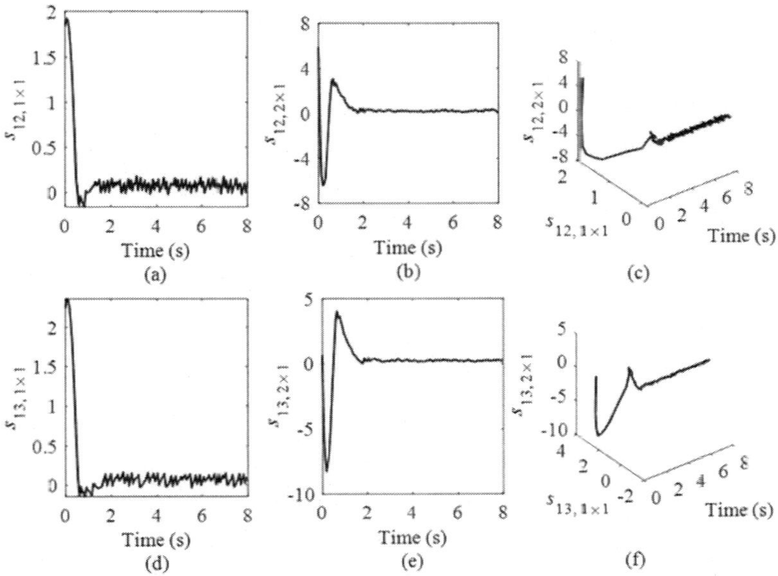

Figure 4.4. Sliding surfaces of the two followers. (a). $s_{12,1\times1}$, (b). $s_{12,2\times1}$, (c). $\mathbf{s}_{12}$, (d). $s_{13,1\times1}$, (e). $s_{13,2\times1}$ and (f). $\mathbf{s}_{13}$.

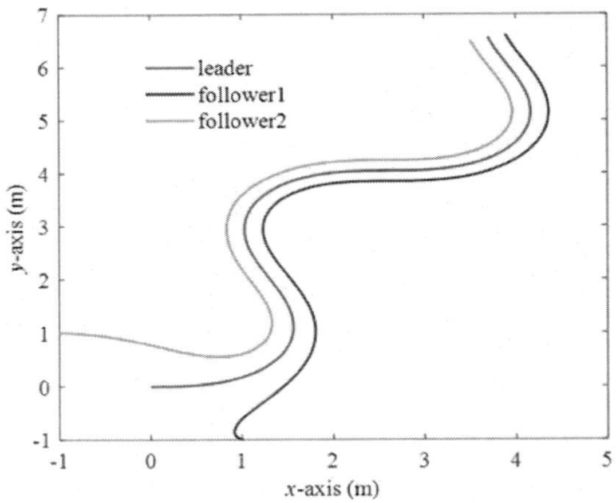

Figure 4.5. Straight line formation of the multi-robot system while moving in a sinusoid.

Figure 4.1 shows the simulation result of the two pairs of robots forming the line formation while moving in a circular trajectory, where the desired linear and angular velocities of the leader robot are set by 0.5m/s and 1rad/s. In Figure 4.1, the two followers follow the leader robot and keep the desired distances and bearing angle from the initial positions.

To demonstrate the control performance of the presented integral sliding mode control method, the results by the presented method are shown in Figures 4.2-4.4, where Figure 4.2 illustrates the system states l_{12}, ψ_{12}, l_{13} and ψ_{13}, Figure 4.3 displays the control inputs of the two followers α_2, β_2, α_3 and β_3. The sliding surface vectors of the two followers are shown in Figure 4.4.

The integral SMC control systems of the two followers are 2 seconds in length to enter the sliding mode stage. In the stage, the trajectories in Figure 4.4c and 4.4f deviate from the sliding surface vectors because of the existence of mismatched uncertainties in the dynamic process. But the integral SMC control laws (4.15) proven in the Theorem 2 guarantees the reachability condition (4.14) is always satisfied. The two sliding surface vectors in Figure 4.4c and 4.4f are locally asymptotically stable after the system states $\mathbf{x}_{12}$ and $\mathbf{x}_{13}$ enter the balls proven in the Theorem 1.

Figure 4.5 illustrates the multi-robot system forms up a straight line when moving along a sinusoid trajectory, where the solid points denote the initial positions of the three mobile robots. In Figure 4.5, the initial location vector for the leader is set by $\mathbf{q}_1 = [0m \;\; 0m \;\; 0rad]^T$ and the initial location vectors of the two followers are set by $\mathbf{q}_2 = [1m \;\; -1m \;\; \frac{3}{4}\pi rad]^T$ and $\mathbf{q}_3 = [1m \;\; -1m \;\; \frac{1}{4}\pi rad]^T$, respectively. According to the above three initial location vectors, the initial distance and orientation of the two followers with respect to the leader in (4.2) are $x_{12} = [\sqrt{2}m \;\; 0m \cdot s^{-1} \;\; \frac{3}{4}\pi rad \;\; 0rad \cdot s^{-1}]^T$ and $x_{13} = [\sqrt{2}m \;\; 0m \cdot s^{-1} \;\; -\frac{1}{4}\pi rad \;\; 0rad \cdot s^{-1}]^T$. The desired system states of the two followers in (4.2) are set by $x_{12} = [0.2m \;\; 0m \cdot s^{-1} \;\; \frac{1}{2}\pi rad \;\; 0rad \cdot s^{-1}]^T$ and $x_{13} = [0.2m \;\; 0m \cdot s^{-1} \;\; -\frac{1}{2}\pi rad \;\; 0rad \cdot s^{-1}]^T$. The acceleration and the angular acceleration of the two followers are subject to $|a_k| \leq \frac{1}{2}\pi m \cdot s^{-2}$ and $|\beta_k| \leq 30rad \cdot s^{-2}$, respectively. The desired linear and angular velocities of the leader robot in Figure 4.5 are set by $v_1 = 1m \cdot s^{-1}$ and $\omega_1 = \sin(t) rad \cdot s^{-2}$. The parameters of the designed integral SMC controllers for the formation maneuver are kept unchanged. They are the same as the parameters for the circle trajectory. Shown in Figure 4.5, the integral SMC method can guarantee both followers effectively follow the leader in presence of mismatched uncertainties although the leader is with its time-variant linear and angular velocities.

CONCLUSION

This paper has addressed the formation control problem for nonholonomic mobile robots with mismatched uncertainties. The formation mechanism is based on the leader-follower scheme. To suppress the mismatched uncertainties when forming up the robots, an integral sliding mode controller is designed. The theoretical analysis in the light of Lyapunov's theory proves the formation control system with mismatched uncertainties is locally asymptotically stable in the sliding mode stage. The presented control method achieves the formation maneuvers of a multi-robot system composed of 3 mobile robots along predefined trajectories, in spite of the presence of mismatched uncertainties. Simulation results have demonstrated the effectiveness, feasibility and robustness of the proposed strategy.

REFERENCES

[1] De La Cruz, C., Carelli, R. (2008). Dynamic model based formation control and obstacle avoidance of multi-robot systems. *Robotica*, 26(3), 345-356.

[2] Antonelli, G., Arrichiello, F., Chiaverini, S. (2010). The NSB control: a behavior-based approach for multi-robot systems. Paladyn, *Journal of Behavioral Robotics*, 1(1), 48-56.

[3] Arrichiello, F., Chiaverini, S., Fossen, T. I. (2006, October). Formation control of underactuated surface vessels using the null-space-based behavioral control. In *Proceedings of 2006 IEEE/RSJ International Conference on Intelligent Robots and Systems* (pp. 5942-5947). IEEE.

[4] Ren, W., Sorensen, N. (2008). Distributed coordination architecture for multi-robot formation control. *Robotics and Autonomous Systems*, 56(4), 324-333.

[5] Hou, S. P., Cheah, C. C. (2010). Can a simple control scheme work for a formation control of multiple autonomous underwater vehicles?. *IEEE Transactions on Control Systems Technology*, 19(5), 1090-1101.

[6] Arai, T., Pagello, E., Parker, L. E. (2002). Advances in multi-robot systems. *IEEE Transactions on robotics and automation*, 18(5), 655-661.

[7] Wang, Q., Wu, M., Huang, Y., Wang, L. (2008). Formation control of heterogeneous multi-robot systems. *IFAC Proceedings Volumes*, 41(2), 6596-6601.

[8] Wang, C., Tnunay, H., Zuo, Z., Lennox, B., Ding, Z. (2018). Fixed-time formation control of multirobot systems: Design and experiments. *IEEE Transactions on Industrial Electronics*, 66(8), 6292-6301.

[9] Tsai, C. C., Wu, H. L., Tai, F. C., Chen, Y. S. (2017). Distributed consensus formation control with collision and obstacle avoidance for uncertain networked omnidirectional multi-robot systems using fuzzy wavelet neural networks. *International Journal of Fuzzy Systems*, 19(5), 1375-1391.

[10] Alonso-Mora, J., Montijano, E., Nägeli, T., Hilliges, O., Schwager, M., Rus, D. (2019). Distributed multi-robot formation control in dynamic environments. *Autonomous Robots*, 43(5), 1079-1100.

[11] Mastellone, S., Stipanović, D. M., Graunke, C. R., Intlekofer, K. A., Spong, M. W. (2008). Formation control and collision avoidance for multi-agent non-holonomic systems: Theory and experiments. *The International Journal of Robotics Research*, 27(1), 107-126.

[12] Freudenthaler, G., Meurer, T. (2020). PDE-based multi-agent formation control using flatness and backstepping: analysis, design and robot experiments. *Automatica*, 115, 108897.

[13] Tran, V. P., Garratt, M., Petersen, I. R. (2020). Switching time-invariant formation control of a collaborative multi-agent system using negative imaginary systems theory. *Control Engineering Practice*, 95, 104245.

[14] Zong, C., Ji, Z., Tian, L., Zhang, Y. (2019). Distributed Multi-Robot Formation Control Based on Bipartite Consensus With Time-Varying Delays. *IEEE Access*, 7, 144790-144798.

[15] Stamouli, C. J., Bechlioulis, C. P., Kyriakopoulos, K. J. (2020). Multi-Agent Formation Control Based on Distributed Estimation With Prescribed Performance. *IEEE Robotics and Automation Letters*, 5(2), 2929-2934.

[16] Wu, H. M., Karkoub, M. (2019). Hierarchical Variable Structure Control for the Path Following and Formation Maintenance of Multi-agent Systems. *Journal of Intelligent & Robotic Systems*, 95(2), 267-277.

[17] Augusto de Holanda, B. M., Madruga, S. P., Brito, A. V., Nascimento, T. P. (2019). Dynamic Leader Allocation in Multi-robot Systems Based on Nonlinear Model Predictive Control. *Journal of Intelligent & Robotic Systems*, 1-18.

[18] Yu, H., Shen, Y., Xia, X. (2013). Adaptive finite-time consensus in multi-agent networks. *Systems & Control Letters*, 62(10), 880-889.

[19] Xie, G., Wang, L. (2009). Moving formation convergence of a group of mobile robots via decentralised information feedback. *International Journal of Systems Science*, 40(10), 1019-1027.

[20] Peng, Z., Wen, G., Rahmani, A. (2013, March). Leader-follower formation control of multiple nonholonomic robots based on backstepping. In *Proceedings of the 28th annual ACM symposium on applied computing* (pp. 211-216).

[21] Guo, J., Qian, D. (2015). Terminal sliding mode formation control of uncertain multiple robots. *International Journal of Advanced Mechatronic Systems*, 6(2-3), 118-127.

[22] Vadali, S. R., Vaddi, S. S., Alfriend, K. T. (2002). An intelligent control concept for formation flying satellites. *International Journal of Robust and Nonlinear Control*, 12(2-3), 97-115.

[23] Huang, C., Zhang, X., Zhang, G. (2019). Improved decentralized finite-time formation control of underactuated USVs via a novel disturbance observer. *Ocean Engineering*, 174, 117-124.

[24] Oh, K. K., Park, M. C., Ahn, H. S. (2015). A survey of multi-agent formation control. *Automatica*, 53, 424-440.

[25] Shtessel, Y., Edwards, C., Fridman, L., Levant, A. (2014). *Sliding mode control and observation* (pp. 1-42). New York: Springer New York.

[26] Utkin, V., Guldner, J., Shi, J. (2017). *Sliding mode control in electro-mechanical systems*. CRC press.

[27] Chang, Y. H., Chen, C. L., Chan, W. S., Lin, H. W. (2013). Type-2 Fuzzy Formation Control for Collision-Free Multi-Robot Systems. *International Journal of Fuzzy Systems*, 15(4).

[28] Zhao, D., Zou, T. (2012). A finite-time approach to formation control of multiple mobile robots with terminal sliding mode. *International Journal of Systems Science*, 43(11), 1998-2014.

[29] Fu, M., Jiao, J., & Yin, S. (2013). Robust coordinated formation for multiple surface vessels based on backstepping sliding mode control. In *Abstract and Applied Analysis* (Vol. 2013). Hindawi.

[30] Defoort, M., Floquet, T., Kokosy, A., Perruquetti, W. (2008). Sliding-mode formation control for cooperative autonomous mobile robots. *IEEE Transactions on Industrial Electronics*, 55(11), 3944-3953.

[31] Park, B. S., Park, J. B., Choi, Y. H. (2011). Robust formation control of electrically driven nonholonomic mobile robots via sliding mode technique. *International Journal of Control, Automation and Systems*, 9(5), 888.

[32] Mi, Y., Fu, Y., Wang, C., Wang, P. (2013). Decentralized sliding mode load frequency control for multi-area power systems. *IEEE Transactions on Power Systems*, 28(4), 4301-4309.

[33] Imani, A., Bahrami, M. (2013). Optimal sliding mode control for spacecraft formation flying in eccentric orbits. *Proceedings of the Institution of Mechanical Engineers, Part I: Journal of Systems and Control Engineering*, 227(5), 474-481.

[34] Sanchez-Torres JD, Loukianov AG, Galicia MI, et al. (2013). Robust nested sliding mode integral control for anti-lock brake system. *Int J Vehicle Des*, 62(2–4), 188-205.

[35] Biglarbegian, M. (2013). A novel robust leader-following control design for mobile robots. *Journal of Intelligent & Robotic Systems*, 71(3-4), 391-402.

In: A Closer Look at Formation Control ISBN: 978-1-53618-177-7
Editor: Dianwei Qian

Chapter 5

LEADER-FOLLOWING FORMATION CONTROL OF MULTIPLE AGENTS VIA DERIVATIVE AND INTEGRAL TERMINAL SLIDING MODE

Dianwei Qian* and Yafei Xi
School of Control and Computer Engineering,
North China Electric Power University, Beijing, China

ABSTRACT

This chapter investigates the formation problem of multiple robots based on the leader-following mechanism. At first, the dynamics of such a leader-follower framework are modeled. The input-output equations are depicted by calculating the relative degree of a leader-following formation system. Furthermore, the derivative and integral terminal sliding mode controller is designed based on the relative degree. Since the formation system suffers from the uncertainties, the nonlinear disturbance observer is adopted to deal with the uncertainties. The stability of the closed-loop control system is proven in the sense of Lynapunov. Finally,

* Corresponding Author's Email: dianwei.qian@ncepu.edu.cn.

some numerical simulations are displayed to verify the feasibility and effectiveness by the designed controller and observer.

Keywords: multiple robots, formation, sliding mode controller, nonlinear disturbance observer, system stability

INTRODUCTION

In recent years, the coordination control scheme of multiple agents system has drawn considerable attention of many researchers in various fields [1, 2]. Multiple agents system is a couple network of many agents, where the agents communicate with each other to achieve some complex duties that are beyond the capacities of single robot [3]. Various control methods have been explored to achieve the coordination of multiple agents, they can be roughly classified into leader-following formation, virtual structure mechanism, graph based approach and behaviour based approach [4, 5].

The leader-following formations are attractive in the coordination control of multi-robot systems [6]. Partly, such formations benefit multiple robots because the formations can have guaranteed formation stability via control design [7]. The basic control idea of the leader-following mechanism is that multiple robots are divided into several leader-following pairs. In the leader-following mechanism, all follower robots share the same leader. In each pair, the leader robot moves along the predefined trajectory while the follower robots track the leader with desired relative distance and angle. In the leader-following system of multiple robots, only partial followers can obtain the state of the leader, and the interaction between follower robots and leader robot is local [8, 9]. Many control methods have been applied in the leader-following multi-robot systems [10-12].

In actuality, it is unavoidable for any robots to be effected by uncertainties such as external disturbances, unmodeled dynamics and parameter perturbations [13, 14]. The dynamics of multi-robot systems becomes uncertain due to these uncertainties. These uncertainties can be categorized into unmatched uncertainties and matched uncertainties [15].

However, the SMC method, as a strong robust tool, has invariant nature to the matched uncertainties when an SMC system enters into the sliding mode [16, 17]. Unfortunately, the effects of unmatched uncertainties cannot be suppressed by the SMC methods [18]. The unmatched uncertainties can challenge the performance of the SMC system seriously. The characteristic of terminal SMC (T-SMC) has its nonlinear sliding surface [19, 20]. Compared with those traditional SMC approaches, the T-SMC method has faster convergence speed and higher accuracy [21, 22]. However, the T-SMC method has the singular problem due to its fractional function. Therefore, the derivative and integral T-SMC (DIT-SMC) method is proposed [23]. The DIT-SMC method is of merit. Due to the existence of the integral term, the sliding mode of the DIT-SMC method starts on the derivative and integral terminal sliding mode surface. Moreover, the DIT-SMC method can guarantee the exact estimation of finite error convergence time, and resolve the singular problem of the T-SMC. On the other hand, the derivative term of the DIT-SMC method can reduce the nonlinear effects to the stability of a DIT-SMC system.

In the previous works, the assumption that the uncertainties have a known boundary is assumed. Concerning the formation maneuvers of multi-robot systems, the assumption is not mild. In fact, the boundary of uncertainties in multiple robots is hard to be known exactly in advance. In case of the lack of the important information, several serious problems may be raised in reality, for example, the decrease of the formation robustness, the deterioration of the formation performance, as far as the deficiency of the formation stability. In order to resolve the problem of the uncertainties, the nonlinear disturbance observer is adopted. The unknown unmatched uncertainties are estimated by the nonlinear disturbance observer. The technique of nonlinear disturbance observer (NDOB) can handle the unmatched uncertainties problem and improve the robustness of the formation control system [24-26].

This chapter deals with the formation problem of multiple robots with uncertainties. The control scheme combining derivative and integral terminal sliding mode and nonlinear disturbance observer is investigated. The derivative and integral terminal sliding mode method allows the system start on the sliding surface. The reaching time of sliding surface is eliminated. The matched uncertainties in formation system are suppressed by the DIT-SMC method. Under the mild assumption that the uncertainties have an unknown boundary, the NDOB is designed to estimate the unmatched uncertainties in the formation system. The estimate errors will convergence to zero in the limited time by setting the parameter of NDOB. In the sense of Lyapunov, the system stability is guaranteed in spite of uncertainties. Finally, some numerical simulations are displayed to illustrate the feasibility and effectiveness.

PROBLEM FORMULATION

Recall the dynamics (1.8). Define $x=\left[x_1\ x_2\ x_3\ x_4\right]^T$, where $x_1=l_{ik}$, $x_2=\dot{l}_{ik}$, $x_3=\psi_{ik}$ and $x_4=\dot{\psi}_{ik}$. Then, the following state equations are deduced from (1.8).

$$\begin{aligned}\dot{x}_{ik}&=f\left(x_{ik},d_{ik}\left(t\right)\right)+B_{ik,1}u_k\\ y_{ik}&=h\left(x_{ik}\right)\end{aligned}\tag{5.1}$$

Here $B_{ik,1}$ is a 2 × 2 matrix whose columns are smooth vector fields. $h\left(x_{ik}\right)$ is the output equation of formation system, and we have

$$f\left(x_{ik},d_{ik}\left(t\right)\right)=A_{ik}x_{ik}+B_{ik,2}d_{ik}\left(t\right)+B_{ik,1}\Delta_k u_k\tag{5.2}$$

In (5.2), A_{ik}, $B_{ik,1}$, $B_{ik,2}$, $h(x_{ik})$ are described by

$$A_{ik}=\begin{bmatrix}0&1&0&0\\0&0&0&0\\0&0&0&1\\0&0&0&0\end{bmatrix},\quad B_{ik,2}=\begin{bmatrix}0&0\\1&0\\0&0\\0&1\end{bmatrix}$$

$$B_{ik,1}=\begin{bmatrix}0 & 0\\ cos\varphi_{ik} & r\sin\varphi_{ik}\\ 0 & 0\\ \dfrac{-\sin\varphi_{ik}}{l_{ik}} & \dfrac{r\cos\varphi_{ik}}{l_{ik}}\end{bmatrix} \text{ and } h(x_{ik})=\begin{bmatrix}x_1\\x_3\end{bmatrix} \tag{5.3}$$

here, $\varphi_{ik}=\psi_{ik}+\theta_{ik}$, $d_{ik}(t)$ denotes the uncertainties of the leader-following formation system (6), written by

$$d_{ik}(t)=L_{ik}(I_2+\Delta_i)u_i+F_{ik}+P_{ik} \tag{5.4}$$

In (5.4), we have

$$L_{ik}=\begin{bmatrix}0 & 0\\ -\cos\psi_{ik} & 0\\ 0 & 0\\ \dfrac{\sin\psi_{ik}}{l_{ik}} & -1\end{bmatrix},\quad F_{ik}=\begin{bmatrix}0\\ \tilde{F}_1\\ 0\\ \tilde{F}_2\end{bmatrix} \text{ and } \quad P_{ik}=\begin{bmatrix}0\\ \tilde{P}_1\\ 0\\ \tilde{P}_2\end{bmatrix} \tag{5.5}$$

here $\tilde{F}_1$, $\tilde{F}_2$, $\tilde{P}_1$ and $\tilde{P}_2$ are defined in (1.8).

CONTROL DESIGN

Relative Degree

Considering the dynamic mode (5.1) of the leader-following formation system, its relative degree can be calculated by

$$L_f h_1(x_{ik}) = \frac{\partial h_1(x_{ik})}{\partial x} f(x_{ik}, d_{ik}) \tag{5.6}$$

$$L_f h_2(x_{ik}) = \frac{\partial h_2(x_{ik})}{\partial x} f(x_{ik}, d_{ik}) \tag{5.7}$$

$$L_{B_{ik,1K}} L_f h_1(x_{ik}, d_{ik}) = L_{B_{ik,1K}}\left(L_f h_1(x_{ik}, d_{ik})\right) = \frac{\partial\left(L_f h_1(x_{ik}, d_{ik})\right)}{\partial x_{ik}} \cdot B_{ik,1K} \tag{5.8}$$

$$L_{B_{ik,1K}} L_f h_2(x_{ik}, d_{ik}) = L_{B_{ik,1K}}\left(L_f h_2(x_{ik}, d_{ik})\right) = \frac{\partial\left(L_f h_2(x_{ik}, d_{ik})\right)}{\partial x_{ik}} \cdot B_{ik,1K} \tag{5.9}$$

here r_K ($K = 1, 2$) is the smallest integer so that the least one of the control input appears in $y_K^{r_K}$ ($K = 1, 2$), then

$$\begin{bmatrix} y_1^{r_1} \\ y_2^{r_2} \end{bmatrix} = \begin{bmatrix} L_f^{r_1} h_1(x_{ik}) + \sum_{K=1}^{m} L_{B_{ik,1}K} L_f h_1(x_{ik}, d_{ik}) u_k \\ L_f^{r_2} h_2(x_{ik}) + \sum_{K=1}^{m} L_{B_{ik,1}K} L_f h_2(x_{ik}, d_{ik}) u_k \end{bmatrix} \tag{5.10}$$

here $r_1 = r_2 = 2$, $m = 2$. $L_{B_{ik,1K}}$, L_f are lie derivatives. Further, the input-output dynamic equation is depicted by

$$\ddot{y}_{ik} = d_{ik} + H(x_{ik}, d_{ik}) u_k \tag{5.11}$$

here $d_{ik} = \tilde{L}_{ik}(I_3 + \Delta_i) u_i + \tilde{P}_{ik} + \tilde{F}_{ik}$, $H(x_{ik}, d_{ik}) = G_{ik}(I_4 + \Delta_k)$. Here I_3, I_4 are 2×2 matrices, u_i is the control input of ith robot, other matrices are described by

$$G_{ik}=\begin{bmatrix}\cos\varphi_{ik} & r\sin\varphi_{ik}\\ -\dfrac{\sin\varphi_{ik}}{l_{ik}} & \dfrac{r\cos\varphi_{k}}{l_{ik}}\end{bmatrix},\quad \tilde{L}_{ik}=\begin{bmatrix}-\cos\psi_{ik} & 0\\ \dfrac{\sin\psi_{ik}}{l_{ik}} & -1\end{bmatrix},\quad \tilde{F}_{ik}=\begin{bmatrix}\tilde{F}_1\\ \tilde{F}_2\end{bmatrix} \text{ and } \tilde{P}_{ik}=\begin{bmatrix}\tilde{P}_1\\ \tilde{P}_2\end{bmatrix} \tag{5.12}$$

Assumption 1: $H(x_{ik}, d_{ik})$ has the normal part G_{ik} and nonlinear part $G_{ik}\Delta_k$, and meets the following inequation

$$\delta_{ik} I \leq H(x_{ik}) H_{dik}^{-1} \leq \delta_{ik}^{-1} I \tag{5.13}$$

here $\delta_{ik}>1$, $H_{dik}=H(x_{ik}^d)$, I is the 2×2 square matrix, x_{ik}^d denotes the desired state vector in (5.1).

$G_{ik}\Delta_k u_k$ is the matched uncertainties in (5.11), meaning the parameter fluctuations of the follower robot k. The term d_{ik} depicts the unmatched uncertainties in the leader-following formation system and consists three parts. Since the formation framework (5.11) is applied to the follower robot, the information of leader robot u_i can not be matched. The terms $\tilde{F}_{ik}$, $\tilde{P}_{ik}$ denote the model uncertainties and external disturbances caused by slipping, friction and obstacles, etc., which are also hard to be matched.

Controller Design

Due to the inherent characteristics of centralization, the scheme is mainly depends on the leader robots and exists the "single point of failure" problem. In order to develop derivative and integral terminal sliding mode approach to coordinate the leader robot i and follower robot k, a recursive structure of the terminal sliding function for high relative-degree MIMO systems (with r_1, $r_2 > 1$) is designed as

$$e_{ik,DO1} = y_{ik,1} - y_{ik,1}^d,\quad e_{ik,DO2} = y_{ik,2} - y_{ik,2}^d \tag{5.14}$$

$$e_{ik,D11} = \dot{e}_{ik,D01}^{P_{11}/q_{11}} + \gamma_{ik,1} e_{ik,DO1} \text{ , } e_{ik,D12} = \dot{e}_{ik,D02}^{P_{12}/q_{12}} + \gamma_{ik,2} e_{ik,DO2} \tag{5.15}$$

$$s_{ik} = \begin{bmatrix} s_{ik,1} \\ s_{ik,2} \end{bmatrix} = \begin{bmatrix} e_{ik,D_{11}} + \lambda_{ik,1} e_{ik,I1} \\ e_{ik,D_{12}} + \lambda_{ik,2} e_{ik,I2} \end{bmatrix} \tag{5.16}$$

where $\dot{e}_{ik,I1} = e_{ik,D_{11}}^{q_{21}/p_{21}}$, $e_{ik,I2} = e_{ik,D_{12}}^{q_{22}/p_{22}}$. $\gamma_{ik,1}$, $\gamma_{ik,2}$, $\lambda_{ik,1}$, $\lambda_{ik,2}$ are all positive constants. $p_{Kj} > q_{Kj}$, here K, j = 1,2. p_{Kj} and q_{Kj} are all odd positive constants.

Theorem 1: Considering the derivative and integral terminal sliding mode surface $s_{ik}(t)$ in (5.16) with the fractional function, the state error of formation system can reach the equilibrium point $e = 0$ at the limited time.

$$T = \max_{K=1,2} \left(\frac{\left| e_{D1K}(0) \right|^{1-q_{2K}/q_{2K}}}{\alpha_K \left(1 - q_{2K} / q_{2K}\right)} + \frac{\left| e_{D0K}(t_{1K}) \right|^{1-q_{1K}/q_{1K}}}{\beta_K \left(1 - q_{1K} / q_{1K}\right)} \right) \tag{5.17}$$

here t_{1K} (K = 1, 2) is the reaching time of terminal slide mode e_{D1K}.

Proof: From (5.16), the sliding mode s_{ik} starts on t = 0. Then the equations $e_{ik,D_{11}} = -\lambda_{ik,1} e_{ik,I1}$ and $e_{ik,D_{12}} = -\lambda_{ik,2} e_{ik,I2}$ can always hold true by control design. Subsequently, substituting $e_{ik,D_{11}} = -\lambda_{ik,1} e_{ik,I1}$, $e_{ik,D_{12}} = -\lambda_{ik,2} e_{ik,I2}$ into $\dot{e}_{ik,I1} = e_{ik,D_{11}}^{q_{21}/p_{21}}$ and $e_{ik,I2} = e_{ik,D_{12}}^{q_{22}/p_{22}}$ respectively yields

$$\dot{e}_{ik,I1}(t) = -\lambda_{ik,1}^{q_{21}/p_{21}} e_{ik,I1}^{q_{21}/p_{21}} \text{ and } \dot{e}_{ik,I2}(t) = -\lambda_{ik,2}^{q_{22}/p_{22}} e_{ik,I2}^{q_{22}/p_{22}} \tag{5.18}$$

The convergent time of sliding mode $e_{ik,I1}$ and $e_{ik,I2}$ can be gotten by solving (5.17).

$$t_{11} = \frac{\left|e_{ik,D11}(0)\right|^{1-q_{21}/p_{21}}}{\lambda_{ik,1}\left(1-q_{21}/p_{21}\right)} \text{ and } t_{12} = \frac{\left|e_{ik,D12}(0)\right|^{1-q_{22}/p_{22}}}{\lambda_{ik,2}\left(1-q_{22}/p_{22}\right)} \tag{5.19}$$

In sliding mode $s_{ik} = 0$, $e_{ik,D_{11}} = -\lambda_{ik,1} e_{ik,I1}$ and $e_{ik,D_{12}} = -\lambda_{ik,2} e_{ik,I2}$ can always hold true. So the reaching time of $e_{ik,D_{11}}$ and $e_{ik,D_{12}}$ are the same as the convergence time of $e_{ik,I1}$ and $e_{ik,I2}$ respectively. When $e_{ik,D_{11}} = e_{ik,D_{12}} = 0$, $e_{ik,I1}$ and $e_{ik,I2}$ will converge to zero successfully. At $t=t_{1K}$ (K=1, 2), $e_{ik,I1}$ and $e_{ik,I2}$ are formulated by

$$\dot{e}_{ik,D01}(t_{11}) = -\gamma_{ik,1}^{q_{11}/p_{11}} e_{ik,D01}^{q_{11}/p_{11}}(t_{11}) \text{ and } \dot{e}_{ik,D02}(t_{12}) = -\gamma_{ik,2}^{q_{12}/p_{12}} e_{ik,D02}^{q_{12}/p_{12}}(t_{12}) \tag{5.20}$$

Solving (5.20), $\dot{e}_{ik,D01}$, $\dot{e}_{ik,D02}$ from $\dot{e}_{ik,D01}(t_{11})$, $\dot{e}_{ik,D02}(t_{12})$ to $\dot{e}_{ik,D01} = \dot{e}_{ik,D01} = 0$ will spend the time

$$t_{01} = \frac{\left|e_{ik,D01}(0)\right|^{1-q_{11}/p_{11}}}{\gamma_{ik,1}\left(1-q_{11}/p_{11}\right)} \text{ and } t_{02} = \frac{\left|e_{ik,D02}(0)\right|^{1-q_{12}/p_{12}}}{\gamma_{ik,2}\left(1-q_{12}/p_{12}\right)} \tag{5.21}$$

Since the sliding mode $s_{ik} = 0$ consists the derivative term and integral term, the time T_K spending from $s_{ik.K} = 0$ (K = 1, 2) to $e_{ik,K} = 0$ (K = 1,2) is the summation of the two terms. Since the fact that each sliding mode $s_{ik,1}$, $s_{ik,2}$ is independent, the time spent for equilibrium point is the maximum of the T_K.

Assumption 2: $\left\|d_{ik}\right\|_\infty \le \bar{d}_{ik}$. It means that the unmatched uncertainties of formation system (5.11) have a boundary.

Differentiating the sliding function s_{ik} with the respect to time t, and substituting (5.11) into the derivative of s_{ik} can get

$$\dot{s}_{ik} = diag\left[\frac{p_{11}}{q_{11}} \dot{e}_{D01}^{(p_{11}/q_{11}-1)}, \frac{p_{12}}{q_{12}} \dot{e}_{D02}^{(p_{12}/q_{12}-1)}\right]\left[d_{ik} + H(x_{ik}, d_{k}) u_k - \ddot{y}_{ik}^{d} + \phi_{ik}\right. \tag{5.22}$$

where $y_{ik}^d = \left[y_{ik,1}^d \quad y_{ik,2}^d \right]^T$ denotes the desired distance and angle between the leader robot and follower robots. $\phi_{ik} = \left[\phi_{ik,1} \quad \phi_{ik,2} \right]^T$. $\phi_{ik,1}$ and $\phi_{ik,2}$ are depicted respectively by

$$\phi_{ik,1} = \gamma_{ik,1}\dot{e}_{D01} + \lambda_{ik,1}e_{D11}^{q_{21}/p_{21}} \text{ and } \phi_{ik,2} = \gamma_{ik,2}\dot{e}_{D02} + \lambda_{ik,2}e_{D12}^{q_{22}/p_{22}} \tag{5.23}$$

The derivative and integral terminal sliding mode control law is set by

$$u = \hat{H}_{dik}^{-1}\left[\ddot{y}_{ik}^d - k_{eik}\left\|\phi_{ik}\right\| diag\left[\frac{q_{11}}{p_{11}}, \frac{q_{12}}{p_{12}} \right] A_{\pi ik}^{-1} \frac{s_{ik}}{\left\|s_{ik}\right\|_1} \right]$$
$$-H_{ik}^{-1}\left(\bar{d}_{ik}(t) - k_{ik}sign(s_{ik}) - \eta_{ik}s_{ik} \right) \tag{5.24}$$

where $sign(s_{ik}) = sign\left[s_{ik,1} \quad s_{ik,2} \right]^T$, k_{eik}, η_{ik} are all the positive constant set by designer. $\bar{d}_{ik}(t)$ is the upper bound of the unmatched uncertainties.

$A_{\pi ik} = \left[A_{\pi ik,1} \quad A_{\pi ik,2} \right]^T$, $A_{\pi ik,K}$ ($K = 1, 2$) are written as

$$A_{\pi ik,K} = \begin{cases} \dot{e}_{ik,0K}^{(p_{1K}/q_{1K}-1)}, for \left| \dot{e}_{ik,0K}^{(p_{1K}/q_{1K}-1)} \right| \geq \varepsilon_{ik} \\ \varepsilon_{ik}, otherwise \end{cases} \tag{5.25}$$

here $\varepsilon_{ik} > 0$, χ_{ik} and k_{eik} have the following conditions.

$$\left\| \left(A_{\pi ik} - diag\left[\dot{e}_{ik,D01}^{(P_{11}/q_{11}-1)}, \dot{e}_{ik,D02}^{(P_{12}/q_{12}-1)} \right] \right) A_{\pi ik}^{-1} \right\| \leq \chi_{ik} < 1 \tag{5.26}$$

$$k_{eik} > 1/(1-\chi_{ik}) \tag{5.27}$$

If a Lyapunov function candicate is picked up as $V_{ik} = \frac{1}{2}s^2$. Substitute the control law (5.24) into (5.22) and consider the conditions (5.25). Then, $\dot{V}_{ik} < 0$ can be guaranteed when $k_{iek} > \bar{d}_{ik}$ holds true. Unfortunately, according to Assumption 2, the unmatched uncertainties d_{ik} in (5.11) are unknown, which means that the upper bound is also unknown. Therefore, it cannot select appropriate parameter k_{iek} to guarantee $\dot{V}_{ik} < 0$. So the stability of control system can not be guaranteed.

DIT-SMC Design Based NDOB

In order to resolve the above problem, the nonlinear disturbance is proposed to estimate the uncertainties d_{ik} in leader-following formation system (5.11). At first, the following assumption is taken into account.

Assumption 3: The unmatched uncertainties possesses a slow change rate, meaning that $\dot{d}_{ik} \approx 0_{2\times 1}$, where $0_{2\times 1} = \begin{bmatrix} 0 & 0 \end{bmatrix}^T$.

Considering the formation dynamic model (5.1), the nonlinear disturbance observer is formulated by

$$\begin{cases} \dot{z}_{ik} = -L_{ik}B_{ik,2}p_{ik} - L_{ik}\left(B_{ik,2}L_{ik}x_{ik} + A_{ik}x_{ik} + B_{ik,1}\left(1+\Delta_k\right)u_k\right) \\ \hat{d}_{ik} = z_{ik} + L_{ik}x_{ik} \end{cases} \tag{5.28}$$

here $z_{ik} \in R^{2\times 1}$, $L_{ik} \in R^{2\times 1}$ and $\hat{d}_{ik} \in R^{2\times 1}$ are the state vector of nonlinear disturbance observer, the observer gain matrix designed by designers and the estimated value of unmatched uncertainties, respectively.

Define the estimate error vector as

$$e_{dik} = d_{ik} - \hat{d}_{ik} \tag{5.29}$$

Differentiate e_{dik} with respect to time t and take Assumption 3 into account. Furthermore the dynamics of e_{dik} is presented by

$$\begin{aligned}
\dot{e}_{dik} &= \dot{d}_{ik} - \dot{\hat{d}}_{ik} \\
&= -\dot{p}_{ik} - L_{ik}\dot{x}_{ik} \\
&= L_{ik}B_{ik,2}p_{ik} + L_{ik}\left(B_{ik,2}L_{ik}x_{ik} + A_{ik}x_{ik} + B_{ik,1}\left(I+\Delta_k\right)u_k\right) \\
&\quad -L_{ik}\left(A_{ik}x_{ik} + B_{ik,1}\left(I+\Delta_k\right)u_k + B_{ik,2}d_{ik}\right) \\
&= L_{ik}B_{ik,2}\left(\hat{d}_{ik} - L_{ik}x_{ik}\right) + L_{ik}\left(B_{ik,2}L_{ik}x_{ik} + A_{ik}x_{ik} + B_{ik,1}\left(I+\Delta_k\right)u_k\right) \\
&\quad -L_{ik}\left(B_{ik,2}d_{ik} + A_{ik}x_{ik} + B_{ik,1}\left(I+\Delta_k\right)u_k\right) \\
&= L_{ik}B_{ik,2}\left(\hat{d}_{ik} - d_{ik}\right) \\
&= -L_{ik}B_{ik,2}e_{dik}
\end{aligned} \tag{5.30}$$

The solution of (5.30) is $e_{dik} = \exp(-L_{ik}B_{ik,2}t)e_{dik}(0)$, which indicates the estimate error will exponentially converge to zero as $t\to\infty$ if $L_{ik}B_{ik,2}$ is set positive constant. Here $e_{dik}(0)$ is the initial state of e_{dik}.

Considering the input-output dynamics (5.11) and the disturbance observer (5.28), the control law based on NDOB is determined by

$$\begin{aligned}
u &= \hat{H}_{dik}^{-1}\left[\ddot{y}_{ik}^{d} - k_{eik}\left\|\phi_{ik}\right\| diag\left[\frac{q_{11}}{p_{11}}, \frac{q_{12}}{p_{12}}\right] A_{\pi ik}^{-1}\frac{s_{ik}}{\left\|s_{ik}\right\|_1}\right] \\
&\quad -H_{ik}^{-1}\left(\hat{d}_{ik}(t) - k_{ik}sign\left(s_{ik}\right) - \eta_{ik}s_{ik}\right)
\end{aligned} \tag{5.31}$$

Theorem 2: Consider the dynamic model of leader-following formation system (5.1), take Assumptions 1, 2 and 3 into account, adopt the input-output model (5.11), design the derivative and integral terminal sliding mode surface (5.16) and the nonlinear disturbance observer (5.28). If the derivative and integral terminal control law is designed by (5.31), then the leader-following formation system with unmatched uncertainties is asymptotically stable when $k_{iek} > e_d^*$.

Proof: Selecting the Lyapunov function as $V_{ik} = \frac{1}{2}s^2$, differentiating V_{ik} with the respect to time t and substituting the $\dot{s}_{ik}$ into the derivative of V_{ik} yields

$$\begin{aligned}\dot{V}_{ik} &= \frac{s_{ik}^T}{\|s_{ik}\|_2} diag\left[\frac{p_{11}}{q_{11}}, \frac{p_{12}}{q_{12}}\right] J_{ik} \begin{Bmatrix} d_{ik} - \hat{d}_{ik} - \eta_{ik} s_{ik} \\ -k_{ik}\,\text{sgn}(s_{ik}) \end{Bmatrix} \\ &+ \frac{s_{ik}^T}{\|s_{ik}\|_2}\phi_{ik} - \delta_{ik} k_{eik} |\phi_{ik}| \|s_{ik}^T H \hat{H}_{dik}^{-1} J_{ik} A_{\pi ik}^{-1} \frac{s_{ik}}{\|s_{ik}\|_1 \|s_{ik}\|_2}\end{aligned} \tag{5.32}$$

here $J_{ik} = diag\left[\dot{e}_{D01}^{(P_{11}/q_{11}-1)}, \dot{e}_{D02}^{(P_{12}/q_{12}-1)}\right]$. Since $p_{1K} > 0$, $q_{1K} > 0$, $p_{2K} > 0$, $q_{2K} > 0$, $p_{1K} > q_{1K}$ and $p_{2k} > q_{2K}$ exist in this controller, $\dot{e}_{D01}^{(P_{11}/q_{11}-1)}$ and $\dot{e}_{D02}^{(P_{12}/q_{12}-1)}$ hold true for all $\dot{e}_{D01}^{(P_{11}/q_{11}-1)} \neq 0$, $\dot{e}_{D02}^{(P_{12}/q_{12}-1)} \neq 0$. According to Assumption 1, the second term and the third term in (5.32) can be deduced from

$$\begin{aligned}& s_{ik}^T \varphi_{ik} - \delta_{ik} k_{eik} \|\phi_{ik}\| s_{ik}^T J_{ik} H \hat{H}_{dik}^{-1} A_{\pi ik}^{-1} \frac{s_{ik}}{\|s_{ik}\|} \\ &\leq \|\phi_{ik}\| \|s_{ik}^T\| - k_{eik} \|\phi_{ik}\| \|s_{ik}^T\| + k_e \|\phi_{ik}\| \|(A_{\pi ik} - J_{ik}) A_{\pi ik}^{-1}\| \|s_{ik}\| \\ &\leq \|\phi_{ik}\| \|s_{ik}\| - k_{eik}(1 - \chi_{ik}) \|\phi_{ik}\| \|s_{ik}\| \\ &\leq 0\end{aligned} \tag{5.33}$$

In (5.33), the condition $k_{eik} > \frac{1}{1 - \chi_{ik}}$ can be picked up so that the $\|\phi_{ik}\| \|s_{ik}\| - k_{eik}(1 - \chi_{ik}) \|\phi_{ik}\| \|s_{ik}\| \leq 0$ holds true.

Further, the first term of $\dot{V}_{ik}$ has the following form of

$$\begin{aligned}
&\frac{s_{ik}^T}{\left\|s_{ik}\right\|_2} diag\left[\frac{p_{11}}{q_{11}}, \frac{p_{12}}{q_{12}}\right] J_{ik}\left\{Z\left(x_{ik}, d_{ik}\right)-\hat{d}_{ik}-k_{ik} \operatorname{sgn}\left(s_{ik}\right)-\eta_{ik} s_{ik}\right\} \\
&\leq \min\left\{diag\left[\frac{p_{11}}{q_{11}}, \frac{p_{12}}{q_{12}}\right] J_{ik}\right\}\left(-k_{ik} \frac{\left\|s_{ik}\right\|_1}{\left\|s_{ik}\right\|_2}-\eta_{ik} \frac{\left\|s_{ik}\right\|_2^2}{\left\|s_{ik}\right\|_2}+\frac{s^T}{\left\|s_{ik}\right\|_2}\begin{bmatrix}\left\|e_{dik}\right\|_\infty \\ \left\|e_{dik}\right\|_\infty\end{bmatrix}\right) \\
&\leq \min\left\{diag\left[\frac{p_{11}}{q_{11}}, \frac{p_{12}}{q_{12}}\right] J_{ik}\right\}\left(-k_{ik} \frac{\left\|s_{ik}\right\|_1}{\left\|s_{ik}\right\|_2}-\eta_{ik} \frac{\left\|s_{ik}\right\|_2^2}{\left\|s_{ik}\right\|_2}+\frac{\left\|s_{ik}\right\|_1}{\left\|s_{ik}\right\|_2}\left\|e_{dik}\right\|_\infty\right) \\
&= \min\left\{diag\left[\frac{p_{11}}{q_{11}}, \frac{p_{12}}{q_{12}}\right] J_{ik}\right\}\left(-\left(k_{ik}-\left\|e_{dik}\right\|_\infty\right) \frac{\left\|s_{ik}\right\|_1}{\left\|s_{ik}\right\|_2}-\eta_{ik} \frac{\left\|s_{ik}\right\|_2^2}{\left\|s_{ik}\right\|_2}\right)
\end{aligned} \tag{5.34}$$

$\eta_{ik} > 0$ and $k_{ik} > \left\|e_{dik}\right\|_\infty$ can be selected in the control design in order that $\frac{s_{ik}^T}{\left\|s_{ik}\right\|_2} diag\left[\frac{p_{11}}{q_{11}}, \frac{p_{12}}{q_{12}}\right] J_{ik}\begin{Bmatrix}Z\left(x_{ik}, d_{ik}\right)-\hat{d}_{ik} \\ -k_{ik} \operatorname{sgn}\left(s_{ik}\right) \\ -\eta_{ik} s_{ik}\end{Bmatrix} \leq 0$ is held true.

$\dot{V}_{ik} < 0$ can be picked up by deducing from (5.32), (5.33) and (5.34). That illustrates the control law can asymptotically stabilize the leader-following formation system by such a derivative and integral terminal sliding mode design. Therefore, the follower robots can trace the leader robot with the desired distance and angle steadily.

The characteristics of DIT-SMC method have: (1) the convergence time T_k can be adjusted by the parameters of the control law; (2) the formation system starts on the derivative and integral terminal sliding surface; (3) the singular problem of T-SMC is avoided; (4) the derivative term can weaken the nonlinear effect.

NUMERICAL SIMULATIONS

Considering the dynamic model of the leader-following formation system (5.1), the derivative and integral terminal sliding mode controller is proposed. There are three robots in the leader-following framework, where the two follower robots track along with the leader robot. The radius of each robot is 0.05m. The parameter fluctuations in formation system are determined by

$$\Delta_i = \Delta_k = \begin{bmatrix} 0.3rand() - 0.15 & 0 \\ 0 & 0.3rand() - 0.15 \end{bmatrix} \tag{5.35}$$

where i = 1 denotes the leader robot, and k = 2, 3 represent the two follower robots. The uncertainties and external disturbances in lumped model (5.1) are depicted by

$$\begin{aligned} \pi_{ix} &= \pi_{iy} = \pi_{i\theta} = 0.5\sin(2\pi t) \\ \pi_{kx} &= \pi_{ky} = \pi_{k\theta} = 0.2\sin(\pi t) \end{aligned} \tag{5.36}$$

The parameters in control design are set as $\gamma_{12,1} = \gamma_{12,2} = \gamma_{13,1} = \gamma_{13,2} = 4$, $\lambda_{12,1} = \lambda_{12,2} = \lambda_{13,1} = \lambda_{13,2} = 1$, $p_{11} = p_{12} = 9$, $q_{11} = q_{12} = 7$, $q_{21} = 3$, $p_{21} = 5$, $q_{22} = 7$, $p_{22} = 9$, $k_{e12} = k_{e13} = 20$, $\varepsilon_{12} = \varepsilon_{13} = 4$, $k_{12} = k_{13} = 2$, $\delta_{12} = \delta_{13} = 4$, $\eta_{12} = \eta_{13} = 0.2$.

Considering the circle trajectory for the formation system in Figure 5.1, the initial state vector are respectively set as x_{12}^0 = [0.5m 0m/s^{-1} 3.2$\pi/4rad$ 0rad/s], x_{13}^0 = [0.707m 0m/s^{-1} 3$\pi/4rad$ 0rad/s]. The desired state vectors are respectively designated as x_{12}^d = [0.13m 0m/s^{-1} $\pi/2rad$ 0rad/s], x_{13}^d= = [0.26m 0m/s^{-1} $\pi/2rad$ 0rad/s]. The desired linear and angular velocities of leader robot is designated as $v_1^d = 0.5m$, $\omega_1^d = 1rad/s$. The simulation results are shown in Figures 5.2-5.4.

Figure 5.1 displays the moving curve of the formation system, which shows the robots are in a line while moving along the circular trajectory. In Figure 5.1, the solid point denotes the initial position of the formation robots. The arrows are the moving directions of the three robots. It is seen form the Figure 5.1 that the follower can track the leader robot with the desired distance and angle while the leader robot is tracking the circular trajectory.

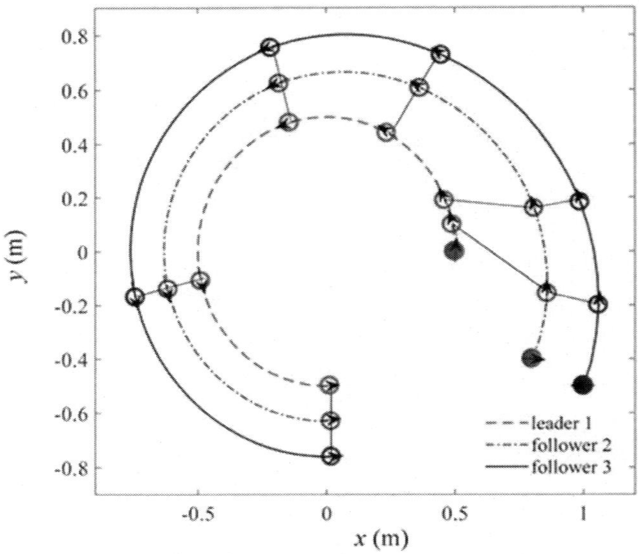

Figure 5.1. Moving trajectory of leader-following formation system.

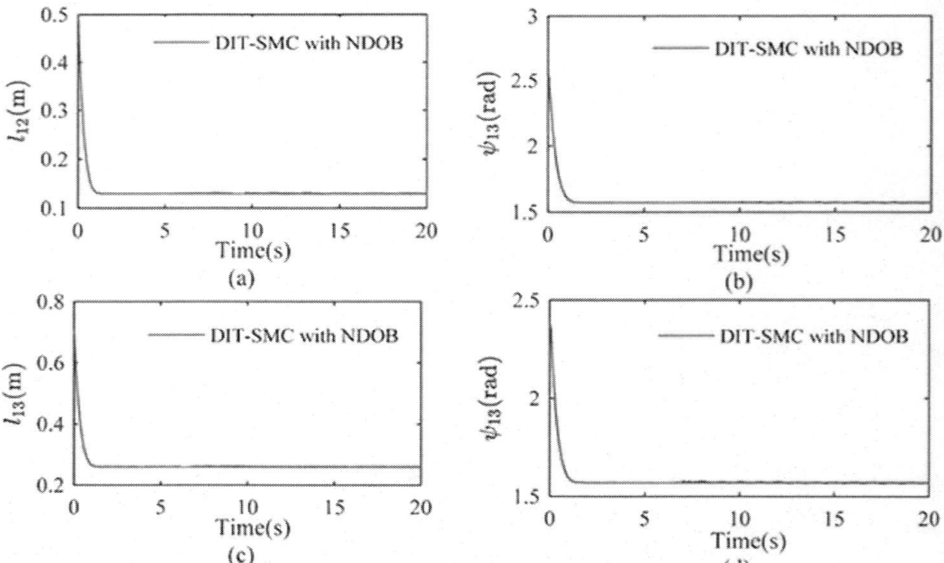

Figure 5.2. Relative distance and angler of leader-following framework in this multi-robot system. a) l_{12}, b) ψ_{12}, c) l_{13} and d) ψ_{13}.

In order to provide more insight into the system performance, the results by the DIT-SMC based NDOB are showed in the Figures 5.2-5.4. The relative distances and relative angles between the leader robot 1 and the two follower robots are displayed in Figure 5.2. Figure 5.3 denotes the control inputs of follower robot 2. Meanwhile, the control inputs of follower robot 3 are shown in the Figure 5.4.

Figure 5.5 denotes the sliding mode vectors of two follower robots. As proven in the thoerom 1, the reaching time of sliding surface will be eliminated and the error of formation system will reach to the equilibrium point in the finite time. From Figure 5.5, the formation system can enter the sliding mode in the beginning, which can guarantee the system stability.

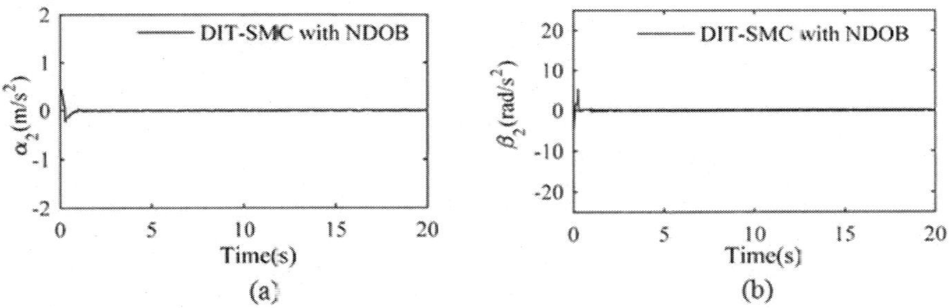

Figure 5.3. Acceleration and angular acceleration of follower robot 2.

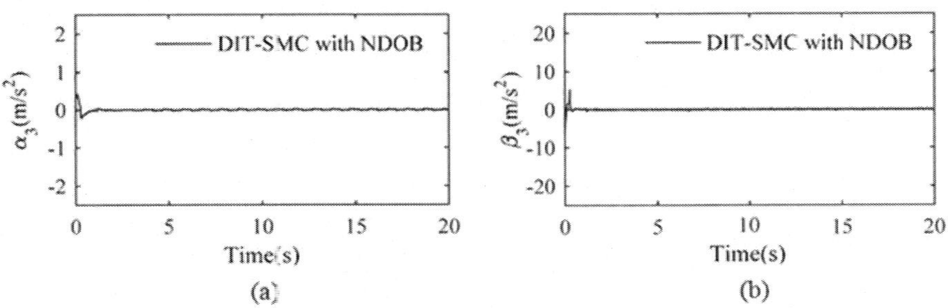

Figure 5.4. Acceleration and angular acceleration of follower robot 3.

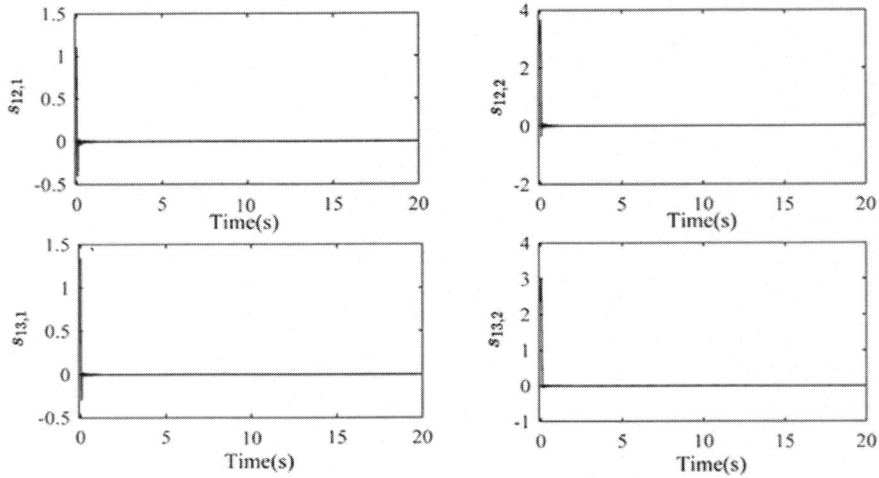

Figure 5.5. Elements of the sliding surfaces.

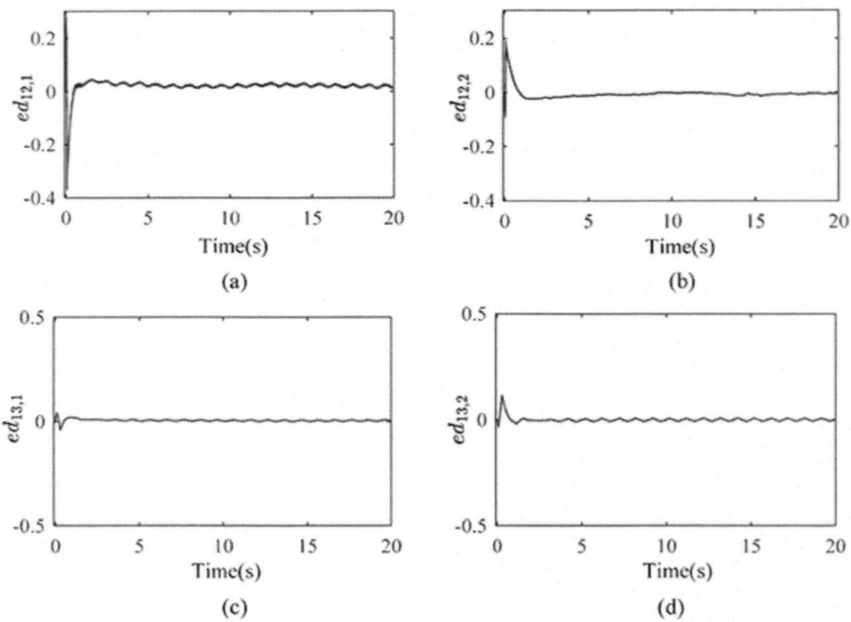

Figure 5.6. Estimate-error (a) $ed_{12,1}$, (b) $ed_{12,2}$, (c) $ed_{13,1}$ and (d) $ed_{13,2}$.

Figure 5.6 illustrates the elements of the estimate-error vectors, where the vectors ed_{12} are showm in the Figure 5.6a and b, and the vectors ed_{13}

are shown in the Figure 5.6c and d. From Figure 5.6, the estimate-error can converfence to zero in the finite time. The value of estimate-error is max when $t = 0$. that is the maxium is less than 0.5. However, the value of k_{ik} is selected 2. So the system stability can be guaranteed.

CONCLUSION

This paper investigates the formation control problem of multi-robot systems based on the leader-follower mechanism. The leader-follower formation system becomes uncertain beceasue of some adverse effects such as the parameter fluctuations, external disturbances and so on. In order to estimate the uncertainties, a control scheme, combining the DIT-SMC and the NDOB, is proposed under the assumption that the uncertainties have an unknown boundary. The stability of the control scheme is proven in the light of Lyapunov theorem. Some simulation results are demonstrated to show the feasibility of the control scheme.

REFERENCES

[1] Qian, D. W., Tong, S. W., Li, C. D. (2017). Observer-based leader-following formation control of uncertain multiple agents by integral sliding mode. *Bulletin of the Polish Academy of Sciences Technical Sciences*, 65(1), 35-44.

[2] Yu, S., Long, X. (2015). Finite-time consensus for second-order multi-agent systems with disturbances by integral sliding mode. *Automatica*, 54. 158-165.

[3] Zhao, D., Zou, T. (2012). A finite-time approach to formation control of multiple mobile robots with terminal sliding mode. *International Journal of Systems Science*, 43(11), 1998-2014.

[4] Qian, D., Xi, Y. (2018). Leader–Follower Formation Maneuvers for Multi-Robot Systems via Derivative and Integral Terminal Sliding Mode. *Applied Sciences*, 8(7), 1045.

[5] Asif, M., Khan, M. J., Memon, A. Y. (2017). Integral terminal sliding mode formation control of non-holonomic robots using leader follower approach. *Robotica*, 35(7), 1473-1487.

[6] Wang, D., Zong, Q., Tian, B., Shao, S., Zhang, X., Zhao, X. (2018). Neural network disturbance observer-based distributed finite-time formation tracking control for multiple unmanned helicopters. *ISA transactions*, 73, 208-226.

[7] Zou, A. M., Kumar, K. D., Hou, Z. G. (2013). Distributed consensus control for multi-agent systems using terminal sliding mode and Chebyshev neural networks. *International Journal of Robust and Nonlinear Control*, 23(3), 334-357.

[8] Liu, H., Cheng, L., Tan, M., Hou, Z. (2014). Containment control of general linear multi-agent systems with multiple dynamic leaders: A fast sliding mode based approach. *IEEE/CAA Journal of Automatica Sinica*, 1(2), 134-140.

[9] Rupp, A., Steinberger, M., Horn, M. (2017). Sliding mode based platooning with non-zero initial spacing errors. *IEEE Control Systems Letters*, 1(2), 274-279.

[10] Qian, D., Li, C., Lee, S., Ma, C. (2017). Robust formation maneuvers through sliding mode for multi-agent systems with uncertainties. *IEEE/CAA Journal of Automatica Sinica*, 5(1), 342-351.

[11] Khanzadeh, A., Pourgholi, M. (2017). Fixed-time leader–follower consensus tracking of second-order multi-agent systems with bounded input uncertainties using non-singular terminal sliding mode technique. *IET Control Theory & Applications*, 12(5), 679-686.

[12] Nair, R. R., Behera, L., Kumar, S. (2017). Event-triggered finite-time integral sliding mode controller for consensus-based formation of multirobot systems with disturbances. *IEEE Transactions on Control Systems Technology*, 27(1), 39-47.

[13] Wu, B., Wang, D., Poh, E. K. (2013). Decentralized sliding-mode control for attitude synchronization in spacecraft formation.

International Journal of Robust and Nonlinear Control, 23(11), 1183-1197.

[14] Zhao, E., Chao, T., Wang, S., Yang, M. (2017). Finite-time formation control for multiple flight vehicles with accurate linearization model. *Aerospace Science and Technology*, 71, 90-98.

[15] Yu, S., Long, X. (2016). Finite-time consensus tracking of perturbed high-order agents with relative information by integral sliding mode. *IEEE Transactions on Circuits and Systems II: Express Briefs*, 63(6), 563-567.

[16] Ma, X., Sun, F., Li, H., He, B. (2017). Neural-network-based integral sliding-mode tracking control of second-order multi-agent systems with unmatched disturbances and completely unknown dynamics. *International Journal of Control, Automation and Systems*, 15(4), 1925-1935.

[17] Tsai, C. C., Yu, C. C., Wu, C. W. (2020). Adaptive distributed BLS-FONTSM formation control for uncertain networking heterogeneous omnidirectional mobile multirobots. *Journal of the Chinese Institute of Engineers*, 43(2), 171-185.

[18] Hou, H., Zhang, Q. (2016). Finite-time synchronization for second-order nonlinear multi-agent system via pinning exponent sliding mode control. *ISA transactions*, 65, 96-108.

[19] Yu, S., Long, X., Guo, G. (2016). Continuous finite-time output consensus tracking of high-order agents with matched and unmatched disturbances. *IET Control Theory & Applications*, 10(14), 1716-1723.

[20] Sinha, A., Mishra, R. K. (2018). Consensus in first order nonlinear heterogeneous multi-agent systems with event-based sliding mode control. *International Journal of Control*, 1-14.

[21] Munir, M., Khan, Q., Deng, F. (2019). Integral Sliding Mode Strategy for Robust Consensus of Networked Higher Order Uncertain Non Linear Systems. *IEEE Access*, 7, 85310-85318.

[22] Chang, C. W., Yang, C. Y., Tao, C. W. (2017). Interval fuzzy sliding-mode formation controller design. *Soft Computing*, 21(14), 4045-4054.

[23] Chiu, C. S. (2012). Derivative and integral terminal sliding mode control for a class of MIMO nonlinear systems. *Automatica*, 48(2), 316-326.

[24] Chen, W. H. (2004). Disturbance observer based control for nonlinear systems. *IEEE/ASME transactions on mechatronics*, 9(4), 706-710.

[25] Mohammadi, A., Tavakoli, M., Marquez, H. J., Hashemzadeh, F. (2013). Nonlinear disturbance observer design for robotic manipulators. *Control Engineering Practice*, 21(3), 253-267.

[26] Yang, J., Li, S., Sun, C., Guo, L. (2013). Nonlinear-disturbance-observer-based robust flight control for airbreathing hypersonic vehicles. *IEEE Transactions on Aerospace and Electronic Systems*, 49(2), 1263-1275.

In: A Closer Look at Formation Control
Editor: Dianwei Qian
ISBN: 978-1-53618-177-7

Chapter 6

EXTREME LEARNING MACHINE-BASED SLIDING MODE FORMATION MANEUVERS FOR MULTI-ROBOT SYSTEMS

***Dianwei Qian*[1,*] *and Jiarong Chen*[1]**
[1]School of Control and Computer Engineering,
North China Electric Power University, Beijing, China

ABSTRACT

This chapter addresses a second-order sliding mode control method for the formation problem of multi-robot systems. This sliding mode control is based on the super-twisting law. In many real-world applications, the robots suffer from a great diversity of uncertainties and disturbances that challenge the super-twisting sliding mode formation maneuvers very much. Especially, such a challenge has the adverse effects on the formation performance when the uncertainties and disturbances have an unknown boundary. This paper focuses on this issue and utilizes the technique of extreme learning machine to meet this challenge. Within the leader-follower framework, this paper investigates the integration of the super-twisting sliding mode control method and the

[*] Corresponding Author's Email: dianwei.qian@ncepu.edu.cn.

extreme learning machine. The output weights of this extreme learning machine are adaptively adjusted so that this integrated formation design has the guaranteed closed-loop stability in the sense of Lyaponov. In the end, some simulations are implemented via a multi-robot platform, which illustrate the superiority and effectiveness of the integrated formation design in spite of uncertainties and disturbances.

Keywords: multi-robot systems, formation maneuvers, 2nd-order sliding mode control, super-twisting algorithm, extreme learning machine

INTRODUCTION

With the emerging of artificial intelligence technologies, multi-robot systems have gained great attention [1, 2]. Such systems not only strengthen and refine the ability of individual robots, but also they provide a platform to display collective behaviors [3]. Compared with a complex robot, multi-robot systems rooted in the real world have broad applications, for example, collaborative projects, military reconnaissance, search and rescue [4-6].

In many industrial, agricultural and maritime situations, multiple robots have to form up some given patterns in order to fulfill a task [7]. In order to manage and coordinate the robots, the formation problem raises up [8]. This problem is originated from some biological phenomena in nature such as schools of fishes in swimming and team of ants in moving. Concerning these biological systems, their formation behaviors exhibit high robustness and hierarchy in that a certain formation mechanism inherently exists [9, 10]. Similarly, the multiple robots call for such a mechanism. Some typical mechanisms have been developed for the robots, that is, the behavior-based algorithm, the virtual structure technique, the leader-follower framework and the artificial potential field approach [11]. From the aspect of control design, the leader-follower framework has visibly blossomed although the mechanism is criticized for its drawback of 'single point of failure' [12, 13]. This paper does not focus on how to design a novel mechanism, but it work at the formation control design.

Consequently, the leader-follower framework is directly adopted because the mechanism is of merit for the small-and-medium scale formation problem.

From the viewpoint of reality, the individual robots of a multi-robot system inevitably are subject to uncertainties and disturbances, which will make the formation dynamics of this multi-robot system uncertain [14-16]. Affected by the adverse factors, the formation control problem of the multiple robots becomes challenging. Many control strategies have been reported [17-20].

As a synthetic tool, the sliding mode control is alternative for the formation problem of uncertain multi-robot systems [21]. So far, some sliding mode-based formation methods have been presented, that is, the first-order sliding mode control, the integral sliding mode control, the derivative and integral terminal sliding mode control and the terminal sliding mode control [23-25]. With regard to the sliding mode control, its most attractive property is the invariance, which can guarantee that a sliding-mode-control system is completely robust despite the matched uncertainties and disturbances.

On the other side, the sliding mode control is also confronted with the dilemma of chattering [26]. As a result, many ideas have been devoted to the decrease and elimination of chattering. Among these ideas, the super-twisting-based sliding mode control technique is advocated because it only needs the information of a sliding mode variable and gets rid of the dependence on the time derivative of this sliding mode variable.

On the assumption that the boundary of uncertainties and disturbances is known, this technique is able to effectively force the sliding mode variable and its time derivative to the origin in finite time [27, 28]. Unfortunately, this assumption is not mild in terms of uncertain multi-robot systems. In reality, one has to overestimate this boundary from the aspect of the closed-loop formation stability. However, the overestimate definitely enlarges the gain of the super-twisting sliding mode control technique. A potential solution is to design a module that can adaptively estimate this boundary.

Motivated by this solution, the Extreme Learning Machine (ELM) is taken into consideration [29-31]. The ELM is a kind of feed-forward neural-networks with the single hidden-layer. The parameters in its hidden layer need no tuning, which are generated randomly and independent of the training data. Compared with the back-propagation algorithm, the training and learning speed of the ELM is much faster. So far, it is reported that the ELM has been successfully applied to microwave filter [32], traffic accident detection [33], air-fuel ratio control [34, 35] and so on. This paper adopts the ELM for the super-twisting sliding mode formation maneuvers of uncertain multi-robot systems. The purpose is to refine the formation performance when the boundary of the uncertainties and disturbances is unknown.

The remainder of this chapter is organized as follows. Section 2 models both a single mobile robot and a leader-follower pair. Section 3 addresses the 2nd-order sliding mode control via the super-twisting algorithm, adopts the ELM to estimate the boundary of the uncertainties and disturbances, and analyzes the closed-loop formation stability in the sense of Lyapunov. In Section 4, we implement the presented control method on a multi-robot system platform. Some numerical results and comparisons are illustrated in Section 4. Finally, conclusions are drawn in Section 5.

DYNAMICS

Recall the dynamics (1.8). Define a vector $\mathbf{x}_{ik} = [x_1\ x_2\ x_3\ x_4]^T$. Let $x_1 = l_{ik}$, $x_2 = \dot{l}_{ik}$, $x_3 = \psi_{ik}$ and $x_4 = \dot{\psi}_{ik}$. According to the formation objective, the relative distance l_{ik} and the relative bearing angle ψ_{ik} are determined as the formation control output. Then, the formation dynamics of this leader-follower pair in the multiple robots can have the form of (6.1) in light of the leader-follower scheme.

$$\begin{aligned}\dot{\mathbf{x}}_{ik} &= \mathbf{f}\left(\mathbf{x}_{ik}, \mathbf{d}_{ik}\right) + \mathbf{g}\left(\mathbf{x}_{ik}, \mathbf{\Delta}_k\right)\mathbf{u}_k \\ \mathbf{y}_{ik} &= \mathbf{h}\left(\mathbf{x}_{ik}\right)\end{aligned} \tag{6.1}$$

here $\mathbf{x}_{ik}$ is the system state vector and $\mathbf{y}_{ik}$ is the system output vector. Further,

$$\begin{aligned}\mathbf{f}\left(\mathbf{x}_{ik}, \mathbf{d}_{ik}\right) &= \mathbf{A}_{ik}\mathbf{x}_{ik} + \mathbf{B}_{ik,2}\mathbf{d}_{ik} \\ \mathbf{g}\left(\mathbf{x}_{ik}, \mathbf{\Delta}_k\right) &= \mathbf{B}_{ik,1} + \mathbf{B}_{ik,1}\mathbf{\Delta}_k\end{aligned} \tag{6.2}$$

In (6.2), $\mathbf{A}_{ik}$, $\mathbf{B}_{ik,1}$, $\mathbf{B}_{ik,2}$ and $\mathbf{h}(\mathbf{x}_{ik})$ are depicted by

$$\mathbf{A}_{ik} = \begin{bmatrix} 0 & 1 & 0 & 0 \\ 0 & 0 & 0 & 0 \\ 0 & 0 & 0 & 1 \\ 0 & 0 & 0 & 0 \end{bmatrix}, \quad \mathbf{B}_{ik,2} = \begin{bmatrix} 0 & 0 \\ 1 & 0 \\ 0 & 0 \\ 0 & 1 \end{bmatrix}$$

$$\mathbf{B}_{ik,1} = \begin{bmatrix} 0 & 0 \\ \cos\varphi_{ik} & r\sin\varphi_{ik} \\ 0 & 0 \\ \frac{-\sin\varphi_{ik}}{l_{ik}} & \frac{r\cos\varphi_{ik}}{l_{ik}} \end{bmatrix}, \quad \mathbf{h}\left(\mathbf{x}_{ik}\right) = \begin{bmatrix} x_1 \\ x_3 \end{bmatrix} \tag{6.3}$$

where $\varphi_{ik} = \psi_{ik} + \theta_{ik}$. $\mathbf{d}_{ik}$ is the lumped term of all the uncertainties and disturbances in the leader-follower pair.

$$\mathbf{d}_{ik} = \mathbf{L}_{ik}\left(\mathbf{I}_2 + \mathbf{\Delta}_i\right)\mathbf{u}_i + \mathbf{F}_{ik} + \mathbf{P}_{ik} \tag{6.4}$$

In (6.4), we have

$$\mathbf{L}_{ik} = \begin{bmatrix} 0 & 0 \\ -\cos\psi_{ik} & 0 \\ 0 & 0 \\ \dfrac{\sin\psi_{ik}}{l_{ik}} & -1 \end{bmatrix}, \quad F_{ik} = \begin{bmatrix} 0 \\ F_1 \\ 0 \\ F_2 \end{bmatrix} \text{ and } P_{ik} = \begin{bmatrix} 0 \\ P_1 \\ 0 \\ P_2 \end{bmatrix} \tag{6.5}$$

here $\mathbf{I}_2$ is a 2×2 identity matrix and F_1, F_2, P_1 and P_2 are written and given in Chapter 1.

The formation dynamics in (6.1) is depicted by the first-order differential equations. Further, the formation dynamics can also be written by the second-order differential equations,

$$\ddot{\mathbb{T}}_{ik} = \mathbb{G}_{ik}\mathbf{u}_k + \mathbb{D}_{ik} \tag{6.6}$$

here

$$\mathbb{T}_{ik} = \begin{bmatrix} l_{ik} & \psi_{ik} \end{bmatrix}^T \quad \mathbb{G}_{ik} = \begin{bmatrix} \cos\varphi_{ik} & r\sin\varphi_{ik} \\ -\dfrac{\sin\varphi_{ik}}{l_{ik}} & \dfrac{r\cos\varphi_{ik}}{l_{ik}} \end{bmatrix}$$

$$\mathbb{D}_{ik} = \begin{bmatrix} d_{ik,1} & d_{ik,2} \end{bmatrix}^T = \mathbb{G}_{ik}\Delta_k\mathbf{u}_k + \mathbb{L}_{ik}(\mathbf{I}_2 + \Delta_i)\mathbf{u}_i + \mathbb{F}_{ik} + \mathbb{P}_{ik}$$

$$\mathbb{L}_{ik} = \begin{bmatrix} -\cos\psi_{ik} & 0 \\ \dfrac{\sin\psi_{ik}}{l_{ik}} & -1 \end{bmatrix}, \quad \mathbb{F}_{ik} = \begin{bmatrix} F_1 \\ F_2 \end{bmatrix}, \quad \mathbb{P}_{ik} = \begin{bmatrix} P_1 \\ P_2 \end{bmatrix} \tag{6.7}$$

Both (6.1) and (6.6) describe the formation dynamics, where (6.1) is in the form of the first-order differential equations and (6.6) has the expressions of the second-order differential equations. Inherently, they are equivalent to each other and both of them can help the following control design.

FORMATION DESIGN

Sliding Surfaces and Input-Output Dynamics

The super-twisting law is a powerful and effective technique that can realize the second-order sliding mode control design. The technique can effectively deal with the controlled plant with a relative degree equal to one with respect to the control input. With regard to the matched uncertainties and disturbances, it can make the sliding mode variable and its time derivative convergent to the origin in finite time. Consequently, we consider this technique as a solution for formation maneuvers of the leader-follower pair. In order to implement the control design, the sliding surfaces, that is, the sliding-mode vector, have to be pre-defined.

$$\mathbf{s}_{ik} = \begin{bmatrix} s_{ik,1} \\ s_{ik,2} \end{bmatrix} = \mathbf{C}_1\left(\begin{bmatrix} l_{ik} \\ \psi_{ik} \end{bmatrix} - \begin{bmatrix} l_{ik}^d \\ \psi_{ik}^d \end{bmatrix}\right) + \mathbf{C}_2\left(\begin{bmatrix} \dot{l}_{ik} \\ \dot{\psi}_{ik} \end{bmatrix} - \begin{bmatrix} \dot{l}_{ik}^d \\ \dot{\psi}_{ik}^d \end{bmatrix}\right) \tag{6.8}$$

here l_{ik}^d and ψ_{ik}^d are the desired relative distance and the desired relative bearing angle of the leader-follower pair, respectively. $\mathbf{C}_1$ and $\mathbf{C}_2$ are 2×2 constant diagonal matrices, given by

$$\mathbf{C}_1 = \begin{bmatrix} c_1 & 0 \\ 0 & c_1 \end{bmatrix} \text{ and } \mathbf{C}_2 = \begin{bmatrix} c_2 & 0 \\ 0 & c_2 \end{bmatrix} \tag{6.9}$$

where both c_1 and c_2 are positive and pre-defined constants.

Differentiate the sliding-mode vector $\mathbf{s}_{ik}$ in (6.8) with the respect to time and substitute the formation dynamics (6.1) into the derivative of $\mathbf{s}_{ik}$. Then, the input–output dynamics are determined by

$$\dot{\mathbf{s}}_{ik} = \frac{\partial \mathbf{s}_{ik}}{\partial t} + \frac{\partial \mathbf{s}_{ik}}{\partial \mathbf{x}_{ik}}\mathbf{f}\left(\mathbf{x}_{ik}, \mathbf{d}_{ik}\right) + \frac{\partial \mathbf{s}_{ik}}{\partial \mathbf{x}_{ik}}\mathbf{g}\left(\mathbf{x}_{ik}, \boldsymbol{\Delta}_k\right)\mathbf{u}_k \tag{6.10}$$

In order to achieve the super-twisting sliding mode control design, the first step is to calculate the relative degree of the dynamics (6.1) with respect to the control input. From (6.8) and (6.10), we have

$$\frac{\partial \mathbf{s}_{ik}}{\partial \mathbf{u}_k} = 0 \quad \text{and} \quad \frac{\partial \dot{\mathbf{s}}_{ik}}{\partial \mathbf{u}_k} = \frac{\partial \mathbf{s}_{ik}}{\partial \mathbf{x}_{ik}} \mathbf{g}\left(\mathbf{x}_{ik}, \mathbf{\Delta}_k\right) \neq 0 \tag{6.11}$$

From (13), it is apparent that the relative degree of $\mathbf{s}_{ik}$ with respect to $\mathbf{u}_k$ is equal to 1. In other word, the super-twisting sliding mode control design is available for formation maneuvers of the multiple robots under the leader-follower scheme.

Let

$$\begin{aligned} \mathbf{a}(\mathbf{x}_{ik}, \mathbf{d}_{ik}, t) &= \frac{\partial \mathbf{s}_{ik}}{\partial t} + \frac{\partial \mathbf{s}_{ik}}{\partial \mathbf{x}_{ik}} \mathbf{f}\left(\mathbf{x}_{ik}, \mathbf{d}_{ik}\right) \\ \mathbf{b}(\mathbf{x}_{ik}, \mathbf{\Delta}_k, t) &= \frac{\partial \mathbf{s}_{ik}}{\partial \mathbf{x}_{ik}} \mathbf{g}\left(\mathbf{x}_{ik}, \mathbf{\Delta}_k\right) \end{aligned} \tag{6.12}$$

Assumption 1: The matrix $\mathbf{b}(\mathbf{x}_{ik}, \mathbf{\Delta}_k, t) \in \Re^{2\times 2}$ in (6.12) contains both known and unknown parts, written by

$$\mathbf{b}(\mathbf{x}_{ik}, \mathbf{\Delta}_k, t) = \mathbf{b}_0(\mathbf{x}_{ik}, t) + \mathbf{b}_1(\mathbf{x}_{ik}, \mathbf{\Delta}_k, t) \tag{6.13}$$

here $\mathbf{b}_0(\mathbf{x}_{ik}, t)$ is a known positively-definite matrix, $\mathbf{b}_1(\mathbf{x}_{ik}, \mathbf{\Delta}_k, t)$ is bounded but unknown and the two parts of $\mathbf{b}(\mathbf{x}_{ik}, \mathbf{\Delta}_k, t)$ satisfy

$$\| \mathbf{b}_1(\mathbf{x}_{ik}, \mathbf{\Delta}_k, t)\mathbf{b}_0^{-1}(\mathbf{x}_{ik}, t) \|_2 < \gamma_1 < 1 \tag{6.14}$$

where γ_1 is a unknown constant.

Assumption 2: The vector $\mathbf{a}(\mathbf{x}_{ik}, \mathbf{d}_{ik}, t) \in \Re^{2\times 1}$ contains both known and unknown parts, depicted by

$$\mathbf{a}(\mathbf{x}_{ik},\mathbf{d}_{ik},t)=\mathbf{a}_0(\mathbf{x}_{ik},t)+\mathbf{a}_1(\mathbf{x}_{ik},\mathbf{d}_{ik},t) \tag{6.15}$$

here $\mathbf{a}_0(\mathbf{x}_{ik},t)$ is a known and bounded vector and $\mathbf{a}_1(\mathbf{x}_{ik},\mathbf{d}_{ik},t)$ is bounded but unknown. Their ∞-norm satisfy

$$\|\mathbf{a}_0(\mathbf{x}_{ik},t)\|_\infty \le \delta_1\sqrt{\|\mathbf{s}_{ik}\|_2} \text{ and } \|\dot{\mathbf{a}}_1(\mathbf{x}_{ik},\mathbf{d}_{ik},t)\|_\infty \le \delta_2 \tag{6.16}$$

In (6.16), here δ_1 and δ_2 are positive but unknown.

Concerning the two assumptions, the input-output dynamics of the sliding-mode vector $\mathbf{s}_{ik}$ in (6.8) can have the form of

$$\dot{\mathbf{s}}_{ik}=\mathbf{a}_0+\mathbf{b}_0\mathbf{u}_k+\mathbf{a}_1+\mathbf{b}_1\mathbf{u}_k \tag{6.17}$$

here $\mathbf{a}_0(\mathbf{x}_{ik},t)$, $\mathbf{a}_1(\mathbf{x}_{ik},\mathbf{d}_{ik},t)$, $\mathbf{b}_0(\mathbf{x}_{ik},t)$ and $\mathbf{b}_1(\mathbf{x}_{ik},\mathbf{\Delta}_k,t)$ are abbreviated by $\mathbf{a}_0$, $\mathbf{a}_1$, $\mathbf{b}_0$ and $\mathbf{b}_1$ for brevity.

Super-Twisting Sliding Mode Control Design

According to the nominal system of (6.17), the super-twisting sliding mode control can be designed as

$$\mathbf{u}_k=\mathbf{b}_0^{-1}(-\mathbf{a}_0+\varpi_k) \tag{6.18}$$

Here

$$\begin{aligned}\varpi_k&=\varpi_{k1}+\varpi_{k2}\\ \varpi_{k1}&=-\alpha_k\sqrt{\|\mathbf{s}_{ik}\|_2}\,\mathrm{sgn}(\mathbf{s}_{ik})\\ \dot{\varpi}_{k2}&=-\chi_k\,\mathrm{sgn}(\mathbf{s}_{ik})\end{aligned} \tag{6.19}$$

In (6.19), α_k and χ_k are positive and they need to be pre-defined. The signum function $\mathrm{sgn}(\mathbf{s}_{ik})$ in (6.19) is defined by $\mathrm{sgn}(\mathbf{s}_{ik}) = \left[\mathrm{sgn}(s_{ik,1}) \quad \mathrm{sgn}(s_{ik,2})\right]^T$.

Select a Lyapunov function candidate

$$V_0 = \left\|\mathbf{s}_{ik}\right\|_2 \tag{6.20}$$

Consider the input-output dynamics (6.6) and substitute (6.14) and (6.15) into (6.17). On Assumptions 1 and 2, the time derivative of (6.20) has the form of

$$\begin{aligned}
\dot{V}_0 &= \frac{\mathbf{s}_{ik}^T \dot{\mathbf{s}}_{ik}}{\left\|\mathbf{s}_{ik}\right\|_2} = \frac{\mathbf{s}_{ik}^T}{\left\|\mathbf{s}_{ik}\right\|_2}(\mathbf{a}_0 + \mathbf{b}_0\mathbf{u}_k + \mathbf{a}_1 + \mathbf{b}_1\mathbf{u}_k) \\
&= \frac{\mathbf{s}_{ik}^T}{\left\|\mathbf{s}_{ik}\right\|_2}\left[\varpi_k + \mathbf{a}_1 + \mathbf{b}_1\mathbf{b}_0^{-1}(-\mathbf{a}_0 + \varpi_k)\right] \\
&= \frac{\mathbf{s}_{ik}^T}{\left\|\mathbf{s}_{ik}\right\|_2}\begin{bmatrix} \left(-\alpha_k - \alpha_k\mathbf{b}_1\mathbf{b}_0^{-1}\right)\sqrt{\|\mathbf{s}_{ik}\|_2}\,\mathrm{sgn}(\mathbf{s}_{ik}) \\ +\left(-\chi_k - \chi_k\mathbf{b}_1\mathbf{b}_0^{-1}\right)\int_0^t \mathrm{sgn}(\mathbf{s}_{ik})\,\mathrm{d}t + \mathbf{a}_1 - \mathbf{b}_1\mathbf{b}_0^{-1}\mathbf{a}_0 \end{bmatrix} \\
&\le \frac{\mathbf{s}_{ik}^T}{\left\|\mathbf{s}_{ik}\right\|_2}\begin{bmatrix} \left(-\alpha_k + \alpha_k\gamma_1\right)\sqrt{\|\mathbf{s}_{ik}\|_2}\,\mathrm{sgn}(\mathbf{s}_{ik}) \\ +\left(-\chi_k + \chi_k\gamma_1\right)\int_0^t \mathrm{sgn}(\mathbf{s}_{ik})\,\mathrm{d}t + \mathbf{a}_1 + \gamma_1\mathbf{a}_0 \end{bmatrix} \\
&= \frac{\left(-\alpha_k + \alpha_k\gamma_1\right)\sqrt{\|\mathbf{s}_{ik}\|_2}}{\left\|\mathbf{s}_{ik}\right\|_2}\mathbf{s}_{ik}^T\,\mathrm{sgn}(\mathbf{s}_{ik}) \\
&\quad + \frac{\left(-\chi_k + \chi_k\gamma_1\right)}{\left\|\mathbf{s}_{ik}\right\|_2}\mathbf{s}_{ik}^T\int_0^t \mathrm{sgn}(\mathbf{s}_{ik})\,\mathrm{d}t + \frac{\mathbf{s}_{ik}^T}{\left\|\mathbf{s}_{ik}\right\|_2}\left(\mathbf{a}_1 + \gamma_1\mathbf{a}_0\right)
\end{aligned} \tag{6.21}$$

Note that the equality exists.

$$\left\|\mathbf{s}_{ik}\right\|_1 = \mathbf{s}_{ik}^T\,\mathrm{sgn}(\mathbf{s}_{ik}) \quad \text{and} \quad \int \mathrm{sgn}(\mathbf{s}_{.})\,\mathrm{d}t = \mathrm{sgn}(\mathbf{s}_{.})\int \mathrm{d}t \tag{6.22}$$

Considering (6.16), (6.21) can be wrtten by

$$\begin{aligned}
\dot{V}_0 &\le \frac{\left(-\alpha_k+\alpha_k\gamma_1\right)\sqrt{\|\mathbf{s}_{ik}\|_2}}{\|\mathbf{s}_{ik}\|_2}\|\mathbf{s}_{ik}\|_1+\frac{\mathbf{s}_{ik}^T}{\|\mathbf{s}_{ik}\|_2}\gamma\mathbf{a}_0 \\
&+\frac{\|\mathbf{s}_{ik}\|_1}{\|\mathbf{s}_{ik}\|_2}\int_0^t\left(-\chi_k-\chi_k\gamma_1\right)dt+\frac{\mathbf{s}_{ik}^T}{\|\mathbf{s}_{ik}\|_2}\int_0^t\dot{\mathbf{a}}_1 dt \\
&\le \frac{\left(-\alpha_k+\alpha_k\gamma_1\right)\sqrt{\|\mathbf{s}_{ik}\|_2}}{\|\mathbf{s}_{ik}\|_2}\|\mathbf{s}_{ik}\|_1+\gamma_1\delta_1\frac{\sqrt{\|\mathbf{s}_{ik}\|_2}}{\|\mathbf{s}_{ik}\|_2}\|\mathbf{s}_{ik}\|_1 \\
&+\frac{\|\mathbf{s}_{ik}\|_1}{\|\mathbf{s}_{ik}\|_2}\int_0^t\left(-\chi_k+\chi_k\gamma_1\right)\mathrm{d}t+\frac{\|\mathbf{s}_{ik}\|_1}{\|\mathbf{s}_{ik}\|_2}\int_0^t\delta_2\,\mathrm{d}t \\
&=\frac{\sqrt{\|\mathbf{s}_{ik}\|_2}\|\mathbf{s}_{ik}\|_1}{\|\mathbf{s}_{ik}\|_2}\left(-\alpha_k+\alpha_k\gamma_1+\gamma_1\delta_1\right)+\frac{\|\mathbf{s}_{ik}\|_1}{\|\mathbf{s}_{ik}\|_2}\int_0^t\left(-\chi_k+\chi_k\gamma_1+\delta_2\right)\mathrm{d}t
\end{aligned} \tag{6.23}$$

Concerning (6.14), $0<\gamma_1<1$. Consequently, one can have $\dot{V}_0<0$ by picking up α_k and χ_k if γ_1, δ_1 and δ_2 are known. Unfortunately, these constants are hardly known in advance, that is, (6.23) theoretically holds true but it is not available in reality. In order to make (6.23) hold true, one possible approach is to overestimate α_k and χ_k so that $\dot{V}_0<0$ can be guaranteed and the closed-loop formation system can have the stability in the sense of Lyapunov.

However, the approach inevitably enlarges the gain of the super-twisting sliding mode control technique, which can definitely have the adverse effects on the formation performance. For this issue, this paper touches the extreme learning machine and fuses it with the super-twisting sliding mode control. Their integration can guarantee the formation stability while the super-twisting sliding mode control technique can have a suitable gain.

Super-Twisting Sliding Mode Control Design via ELM

The ELM is a learning algorithm of single hidden layer feed-forward networks. By the algorithm, the input weights are randomly chosen, the hidden-layer biases are randomly assigned and the output weights are analytically determined. The reason that the gain of the super-twisting sliding mode is overestimated is that some boundaries are unknown. With the help of the ELM, one possible approach is to estimate these uncertainties and disturbance online, which can avoid the drawback of overestimating the boundaries.

Considering the formation dynamics in the form of the second-order differential equations (6.6), the sliding surfaces (6.8) can be written by

$$\mathbf{s}_{ik} = \mathbf{C}_1\left(\mathbb{T}_{ik} - \mathbb{T}_{ik}^d\right) + \mathbf{C}_2\left(\dot{\mathbb{T}}_{ik} - \dot{\mathbb{T}}_{ik}^d\right) \tag{6.24}$$

Then, the time derivative of (6.24) is determined by

$$\dot{\mathbf{s}}_{ik} = \mathbf{C}_1\left(\dot{\mathbb{T}}_{ik} - \dot{\mathbb{T}}_{ik}^d\right) + \mathbf{C}_2\left(\ddot{\mathbb{T}}_{ik} - \ddot{\mathbb{T}}_{ik}^d\right) \tag{6.25}$$

For the formation maneuvers with constant l_{ik}^d and ψ_{ik}^d, we have

$$\dot{\mathbb{T}}_{ik}^d = \ddot{\mathbb{T}}_{ik}^d = \begin{bmatrix} 0 & 0 \end{bmatrix}^T \tag{6.26}$$

Consequently, (6.25) has the form of

$$\dot{\mathbf{s}}_{ik} = \mathbf{C}_1\dot{\mathbb{T}}_{ik} + \mathbf{C}_2\ddot{\mathbb{T}}_{ik} \tag{6.27}$$

Substituting (6.6) into (6.27) yields

$$\dot{\mathbf{s}}_{ik} = \mathbf{C}_1\dot{\mathbb{T}}_{ik} + \mathbf{C}_2\mathbb{G}_{ik}\mathbf{u}_k + \mathbf{C}_2\mathbb{D}_{ik} \tag{6.28}$$

Let

$$\dot{\mathbf{s}}_{ik} = \mathbf{C}_1\dot{\mathbb{T}}_{ik} + \mathbf{C}_2\mathbb{G}_{ik}\mathbf{u}_k + \mathbf{C}_2\mathbb{D}_{ik} = \varpi_k \tag{6.29}$$

where ϖ_k is defined in (21). Then, $\mathbf{u}_k$ can be obtained by

$$\mathbf{u}_k = \left(\mathbf{C}_2 \mathbb{G}_{ik}\right)^{-1}\left(-\mathbf{C}\,\dot{\mathbb{T}}_{ik} - \mathbf{C}_2 \mathbb{D}_{ik} + \varpi_k\right) \tag{6.30}$$

From (6.6), $\mathbb{D}_{ik}$ contains all the uncertainties and disturbances. Here the ELM is designed to estimate $\mathbb{D}_{ik}$ in real time. Replacing $\mathbb{D}_{ik}$ by it estimate $\hat{\mathbb{D}}_{ik}$ in (6.30) yields

$$\mathbf{u}_k = \left(\mathbf{C}_2 \mathbb{G}_{ik}\right)^{-1}\left(-\mathbf{C}\,\dot{\mathbb{T}}_{ik} - \mathbf{C}_2 \hat{\mathbb{D}}_{ik} + \varpi_k\right) \tag{6.31}$$

In (6.4), $\mathbb{D}_{ik}$ is the function of some variables, where φ_{ik}, l_{ik} and ψ_{ik} play an important role. Here the three variables are chosen as the input nodes of the ELM, that is, $\mathbf{z} \in \Re^{3\times1}$ is the input vector and $\mathbf{z} = \left[l_{ik} \quad \psi_{ik} \quad \varphi_{ik}\right]^T$. Without doubt, the output vector is just $\hat{\mathbb{D}}_{ik} \in \Re^{2\times1}$, namely, there are two output nodes located at the output layer.

Assign M hidden nodes as the hidden layer. Then, the weights between the input and hidden layers can be defined by $\mathbf{w} \in \Re^{M\times3}$. The input bias vector of the hidden nodes is defined by $\mathbf{c} \in \Re^{M\times1}$. The activation function of the hidden layer is selected as a sigmoidal function. Then, the output of the lth $(l = 1, 2, \cdots, M)$ hidden-layer node can be calculated by

$$h_l(\mathbf{w}_l \cdot \mathbf{z} + c_l) = \frac{1}{1 + e^{-(\mathbf{w}_l \cdot \mathbf{z} + c_l)}} \tag{6.32}$$

Here $\mathbf{w}_l \in \Re^{1\times3}$ is the lth row of $\mathbf{w}$ and c_l is the lth element of $\mathbf{c}_l$.

The output weights between the hidden-layer nodes and the output-layer nodes are defined by $\Theta = \left[\Theta_1 \quad \Theta_2\right] \in \Re^{M\times2}$, $\Theta_1 \in \Re^{M\times1}$ and $\Theta_2 \in \Re^{M\times1}$. Finally, the output vector of the ELM can be calculated by

$$\hat{\mathbb{D}}_{ik} = \Theta^T \mathbb{H} \tag{6.33}$$

Here $\mathbb{H} = [h_1 \quad h_2 \quad \cdots \quad h_l \quad \cdots \quad h_M]^T$.

According to the universal approximation theorem of the single hidden layer feed-forward networks, there exists the optimal output weights $\Theta^* \in \Re^{M\times 2}$ to approximate $\mathbb{D}_{ik}$ so that we have

$$\mathbb{D}_{ik} = \Theta^{*T}\mathbb{H} + \varepsilon(\mathbf{z}) \tag{6.34}$$

Here $\varepsilon(\mathbf{z}) \in \Re^{2\times 1}$ is an approximation error vector and it can be arbitrarily reduced by increasing the number of hidden-layer nodes. Therefore, it is assumed that

$$\left\|\varepsilon(\mathbf{z})\right\|_\infty < \bar{\varepsilon} \tag{6.35}$$

where $\bar{\varepsilon}$ is an arbitrary small constant.

Theorem 1: Consider the formation dynamics (6.1) and (6.6) on the assumptions 1 and 2. Utilize the super-twisting sliding mode control (6.31).

Design the ELM to estimate $\hat{\mathbb{D}}_{ik}$ online by (6.33). If the weights between the hidden and output layers of the ELM are adjusted by (6.36), then the formation control system is of asymptotically stable.

$$\dot{\Theta}_q = \eta \frac{c_2}{\left\|\mathbf{s}_{ik}\right\|_2} s_{ik,q}\mathbb{H} \quad q = 1,2 \tag{6.36}$$

where η is a positive constant, both c_2 and $s_{ik,q}$ are defined by (6.8).

Proof: Take the following Lyapunov function candidate into consideration.

$$V = \left\|\mathbf{s}_{ik}\right\|_2 + \frac{1}{2\eta}\sum_{q=1}^{2}\tilde{\Theta}_q^T\tilde{\Theta}_q \tag{6.37}$$

Here $\tilde{\Theta} = \left[\tilde{\Theta}_1 \quad \tilde{\Theta}_2 \right]$ is defined by

$$\tilde{\Theta} = \Theta^* - \Theta \tag{6.38}$$

The time derivative of V can have the form of

$$\dot{V} = \frac{\mathbf{s}_{ik}^T \dot{\mathbf{s}}_{ik}}{\left\| \mathbf{s}_{ik} \right\|_2} + \frac{1}{\eta} \sum_{q=1}^{2} \tilde{\Theta}_q^T \dot{\tilde{\Theta}}_q \tag{6.39}$$

Substituting (6.28) into (6.39) yields

$$\dot{V} = \frac{\mathbf{s}_{ik}^T}{\left\| \mathbf{s}_{ik} \right\|_2} (\mathbf{C}_1 \dot{\mathbb{T}}_{ik} + \mathbf{C}_2 \mathbb{G}_{ik} \mathbf{u}_k + \mathbf{C}_2 \mathbb{D}_{ik}) + \frac{1}{\eta} \sum_{q=1}^{2} \tilde{\Theta}_q^T \dot{\tilde{\Theta}}_q \tag{6.40}$$

Then, according to the designed formation control (6.31), we have

$$\begin{aligned}
\dot{V} &= \frac{\mathbf{s}_{ik}^T}{\left\| \mathbf{s}_{ik} \right\|_2} \left(\mathbf{C}_2 \mathbb{D}_{ik} - \mathbf{C}_2 \hat{\mathbb{D}}_{ik} + \varpi_k \right) + \frac{1}{\eta} \sum_{q=1}^{2} \tilde{\Theta}_q^T \dot{\tilde{\Theta}}_q \\
&= \frac{\mathbf{s}_{ik}^T}{\left\| \mathbf{s}_{ik} \right\|_2} \left[\mathbf{C}_2 \left(\Theta^{*T} \mathbb{H} - \varepsilon(\mathbf{z}) \right) - \mathbf{C}_2 \left(\Theta^T \mathbb{H} \right) + \varpi_k \right] + \frac{1}{\eta} \sum_{q=1}^{2} \tilde{\Theta}_q^T \dot{\tilde{\Theta}}_q \\
&= \frac{\mathbf{s}_{ik}^T}{\left\| \mathbf{s}_{ik} \right\|_2} \left[\mathbf{C}_2 \left(\Theta^{*T} - \Theta^T \right) \mathbb{H} - \mathbf{C}_2 \varepsilon(\mathbf{z}) + \varpi_k \right] - \frac{1}{\eta} \sum_{q=1}^{2} \tilde{\Theta}_q^T \dot{\tilde{\Theta}}_q \\
&= \frac{\mathbf{s}_{ik}^T}{\left\| \mathbf{s}_{ik} \right\|_2} \left[\mathbf{C}_2 \tilde{\Theta}^T \mathbb{H} - \mathbf{C}_2 \varepsilon(\mathbf{z}) + \varpi_k \right] + \frac{1}{\eta} \sum_{q=1}^{2} \tilde{\Theta}_q^T \dot{\tilde{\Theta}}_q
\end{aligned} \tag{6.41}$$

From (6.19), we can obtain

$$\begin{aligned}\dot{V} &= \frac{\mathbf{s}_{ik}^T}{\left\|\mathbf{s}_{ik}\right\|_2}\left[\mathbf{C}_2\tilde{\Theta}^T\mathbb{H} - \mathbf{C}_2\varepsilon(\mathbf{z}) - \alpha_k\sqrt{\|\mathbf{s}_{ik}\|_2}\,\mathrm{sgn}(\mathbf{s}_{ik}) - \int_0^t \chi_k\,\mathrm{sgn}(\mathbf{s}_{ik})\,\mathrm{d}t\right] \\ &\quad + \frac{1}{\eta}\sum_{q=1}^{2}\tilde{\Theta}_q^T\dot{\tilde{\Theta}}_q \\ &= \frac{\mathbf{s}_{ik}^T}{\left\|\mathbf{s}_{ik}\right\|_2}\mathbf{C}_2\tilde{\Theta}^T\mathbb{H} + \frac{1}{\eta}\sum_{q=1}^{2}\tilde{\Theta}_q^T\dot{\tilde{\Theta}}_q - \frac{\mathbf{s}_{ik}^T}{\left\|\mathbf{s}_{ik}\right\|_2}\mathbf{C}_2\varepsilon(\mathbf{z}) \\ &\quad - \frac{\mathbf{s}_{ik}^T}{\left\|\mathbf{s}_{ik}\right\|_2}\left[\alpha_k\sqrt{\|\mathbf{s}_{ik}\|_2} + \chi_k\int_0^t \mathrm{d}t\right]\mathrm{sgn}(\mathbf{s}_{ik}) \\ &= \frac{\mathbf{s}_{ik}^T}{\left\|\mathbf{s}_{ik}\right\|_2}\mathbf{C}_2\begin{bmatrix}\tilde{\Theta}_1^T\mathbb{H}\\ \tilde{\Theta}_2^T\mathbb{H}\end{bmatrix} + \frac{1}{\eta}\sum_{q=1}^{2}\tilde{\Theta}_q^T\dot{\tilde{\Theta}}_q - \frac{\mathbf{s}_{ik}^T}{\left\|\mathbf{s}_{ik}\right\|_2}\mathbf{C}_2\varepsilon(\mathbf{z}) \\ &\quad - \frac{\mathbf{s}_{ik}^T}{\left\|\mathbf{s}_{ik}\right\|_2}\left[\alpha_k\sqrt{\|\mathbf{s}_{ik}\|_2} + \chi_k\int_0^t \mathrm{d}t\right]\mathrm{sgn}(\mathbf{s}_{ik})\end{aligned} \tag{6.42}$$

Since $\mathbf{C}_2$ is a diagonal matrix, (6.42) can be written by

$$\begin{aligned}\dot{V} &= \frac{\mathbf{s}_{ik}^T}{\left\|\mathbf{s}_{ik}\right\|_2}\mathbf{C}_2\begin{bmatrix}\tilde{\Theta}_1^T\mathbb{H}\\ \tilde{\Theta}_2^T\mathbb{H}\end{bmatrix} + \frac{1}{\eta}\sum_{q=1}^{2}\tilde{\Theta}_q^T\dot{\tilde{\Theta}}_q - \frac{\mathbf{s}_{ik}^T}{\left\|\mathbf{s}_{ik}\right\|_2}\mathbf{C}_2\varepsilon(\mathbf{z}) \\ &\quad - \frac{\mathbf{s}_{ik}^T}{\left\|\mathbf{s}_{ik}\right\|_2}\left[\alpha_k\sqrt{\|\mathbf{s}_{ik}\|_2} + \chi_k\int_0^t \mathrm{d}t\right]\mathrm{sgn}(\mathbf{s}_{ik}) \\ &= c_2\frac{\mathbf{s}_{ik}^T}{\left\|\mathbf{s}_{ik}\right\|_2}\begin{bmatrix}\tilde{\Theta}_1^T\mathbb{H}\\ \tilde{\Theta}_2^T\mathbb{H}\end{bmatrix} + \frac{1}{\eta}\sum_{q=1}^{2}\tilde{\Theta}_q^T\dot{\tilde{\Theta}}_q - \frac{\mathbf{s}_{ik}^T}{\left\|\mathbf{s}_{ik}\right\|_2}\mathbf{C}_2\varepsilon(\mathbf{z}) \\ &\quad - \frac{\mathbf{s}_{ik}^T}{\left\|\mathbf{s}_{ik}\right\|_2}\left[\alpha_k\sqrt{\|\mathbf{s}_{ik}\|_2} + \chi_k\int_0^t \mathrm{d}t\right]\mathrm{sgn}(\mathbf{s}_{ik}) \\ &= \frac{c_2}{\left\|\mathbf{s}_{ik}\right\|_2}\sum_{q=1}^{2}s_{ik,q}\tilde{\Theta}_q^T\mathbb{H} + \frac{1}{\eta}\sum_{q=1}^{2}\tilde{\Theta}_q^T\dot{\tilde{\Theta}}_q - \frac{\mathbf{s}_{ik}^T}{\left\|\mathbf{s}_{ik}\right\|_2}\mathbf{C}_2\varepsilon(\mathbf{z}) \\ &\quad - \frac{\mathbf{s}_{ik}^T}{\left\|\mathbf{s}_{ik}\right\|_2}\left[\alpha_k\sqrt{\|\mathbf{s}_{ik}\|_2} + \chi_k\int_0^t \mathrm{d}t\right]\mathrm{sgn}(\mathbf{s}_{ik})\end{aligned} \tag{6.43}$$

Consider the condition in (6.36) and the definition in (6.38). We can have

$$\dot{\tilde{\Theta}}_q = -\dot{\Theta}_q = -\eta\frac{c_2}{\left\|\mathbf{s}_{ik}\right\|_2}s_{ik,q}\mathbb{H} \tag{6.44}$$

Then, (6.43) becomes

$$\begin{aligned}\dot{V} &= -\frac{\mathbf{s}_{ik}^T}{\|\mathbf{s}_{ik}\|_2}\mathbf{C}_2\varepsilon(\mathbf{z}) - \frac{\mathbf{s}_{ik}^T}{\|\mathbf{s}_{ik}\|_2}\left[\alpha_k\sqrt{\|\mathbf{s}_{ik}\|_2} + \chi_k\int_0^t \mathrm{d}t\right]\mathrm{sgn}(\mathbf{s}_{ik}) \\ &= -c_2\frac{\mathbf{s}_{ik}^T}{\|\mathbf{s}_{ik}\|_2}\varepsilon(\mathbf{z}) - \left[\alpha_k\sqrt{\|\mathbf{s}_{ik}\|_2} + \chi_k\int_0^t \mathrm{d}t\right]\frac{\|\mathbf{s}_{ik}\|_1}{\|\mathbf{s}_{ik}\|_2} \\ &\le c_2\bar{\varepsilon}\frac{\|\mathbf{s}_{ik}\|_1}{\|\mathbf{s}_{ik}\|_2} - \left[\alpha_k\sqrt{\|\mathbf{s}_{ik}\|_2} + \chi_k\int_0^t \mathrm{d}t\right]\frac{\|\mathbf{s}_{ik}\|_1}{\|\mathbf{s}_{ik}\|_2}\end{aligned} \tag{6.45}$$

From (6.45), we can select some suitable α_k and χ_k to make $\dot{V} < 0$ so that the formation system becomes asymptotically stable.

Although $\bar{\varepsilon}$ in (6.45) is still unknown in advance, the universal approximation theorem of the single hidden layer feed-forward networks (6.35) indicates $\bar{\varepsilon}$ can be an arbitrary small constant. This fact avoid the drawback of overestimating α_k and χ_k in (6.18) from the aspect of the formation stability. Thus, the super-twisting sliding mode formation maneuvers via ELM can contribute to the improvement of the formation performance.

NUMERICAL RESULTS

Multi-Robot Platform

This section will integrate the super-twisting sliding mode control and the ELM for an uncertain multi-robot system. In order to verify the designed method, we only consider the small-scale formation so that a multi-robot simulation platform with three mobile robots is taken into consideration. The Robot 1 acts as the leader and there are 2 leader-follower pairs in this platform. The Robot 2 and the Robot 3 act as the followers and they are coordinated by the leader. In the small-scale multi-robot system, some assumptions such as no collisions and no communication delay can easily be held true so that we can focus on the formation control design.

The robots' radius r is set to 0.05m. The uncertainties of the individual robots are set as follows.

$$\Delta_1 = \Delta_2 = \Delta_3 = \begin{bmatrix} \bar{\Delta} & 0 \\ 0 & \bar{\Delta} \end{bmatrix} \tag{6.46}$$

Here $\bar{\Delta}$ is set by 0.3×rad−0.2 and rad is a random number between 0 and 1. Since $\bar{\Delta}$ is related to the changes of the robots' physical parameters, here we consider that they are the same as each other.

In (6.1), the terms $\pi_{ix} - \pi_{kx}$ and $\pi_{iy} - \pi_{ky}$ exist. It is not representative to define $\boldsymbol{\pi}_i = \boldsymbol{\pi}_k$.

Therefore, they are set as (6.47) in the multi-robot platform.

$$\begin{aligned} \pi_{1x} &= \pi_{1y} = \pi_{1\theta} = 0.5\sin(2\pi t) \\ \pi_{2x} &= \pi_{2y} = \pi_{2\theta} = 0.3\sin(2\pi t) \\ \pi_{3x} &= \pi_{3y} = \pi_{3\theta} = 0.3\sin(2\pi t) \end{aligned} \tag{6.47}$$

Without loss of generality, the parameters concerning the super-twisting sliding mode control and the ELM are the same as each other for the two followers. With regard to the super-twisting sliding mode control, its parameters are determined by $\alpha_2 = \alpha_3 = 8$ and $\chi_2 = \chi_3 = 17$. $\mathbf{C}_1$ and $\mathbf{C}_2$ in the sliding surfaces are set by

$$\mathbf{C}_1 = \begin{bmatrix} 800 & 0 \\ 0 & 800 \end{bmatrix} \text{ and } \mathbf{C}_2 = \begin{bmatrix} 56 & 0 \\ 0 & 56 \end{bmatrix} \tag{6.48}$$

As far as the ELM is concerned, $\eta = 1000$, the number of the hidden-layer nodes is set to $M = 500$, the weights between the input and hidden layers are random in the closed interval $[-1 \quad 1]$ and the biases of the hidden nodes are random in the closed interval $[0 \quad 1]$.

Formation Tasks

In Figure 6.1, the multi-robot platform carries out the task of string formation when moving along a circular trajectory, where the red means the leader robot and the green and blue delegate the two followers. The initial postures of the three robots are allocated at

$$\begin{aligned}
\mathbf{q}_1 &= [0.5\text{m} \quad 0\text{m} \quad 0.5\pi\text{rad}]^T \\
\mathbf{q}_2 &= [0.8\text{m} \quad -0.4\text{m} \quad 0\text{rad}]^T \\
\mathbf{q}_3 &= [1.1\text{m} \quad -0.5\text{m} \quad \pi\text{rad}]^T
\end{aligned} \tag{6.49}$$

According to the initial postures (6.49) and the formation task, the initial states of the formation dynamics (6.1) can be calculated as

$$\begin{aligned}
\mathbf{x}_{12}(0) &= [0.48\text{m} \quad 0\text{m/s} \quad 1.24\pi\text{rad} \quad 0\text{rad/s}]^T \\
\mathbf{x}_{13}(0) &= [0.74\text{m} \quad 0\text{m/s} \quad 1.26\pi\text{rad} \quad 0\text{rad/s}]^T
\end{aligned} \tag{6.50}$$

With regard to the formation task, the desired states can be determined by

$$\begin{aligned}
\mathbf{x}_{12}^d &= [0.13\text{m} \quad 0\text{m/s} \quad -0.5\pi\text{rad} \quad 0\text{rad/s}]^T \\
\mathbf{x}_{13}^d &= [0.26\text{m} \quad 0\text{m/s} \quad -0.5\pi\text{rad} \quad 0\text{rad/s}]^T
\end{aligned} \tag{6.51}$$

Figure 6.2 demonstrates the state variables when the multi-robot system fulfills the formation task in Figure 6.1. From Figure 6.2, the presented method (short for STW with ELM) can improve the performance of the system state variables.

From this aspect, the ELM can benefit the improvement of the control performance. Furthermore, the control inputs of the three control methods applied to the follower 2 and the follower 3 are illustrated in Figures 6.3 and 6.4, respectively.

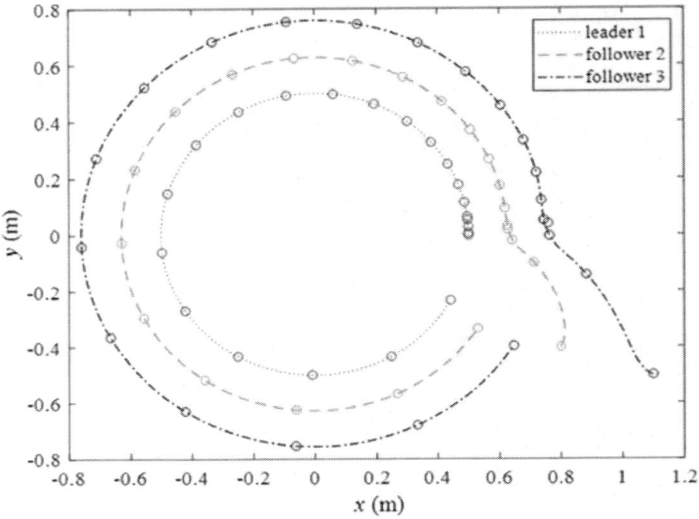

Figure 6.1. String formation of this multi-robot platform when moving along a circular trajectory.

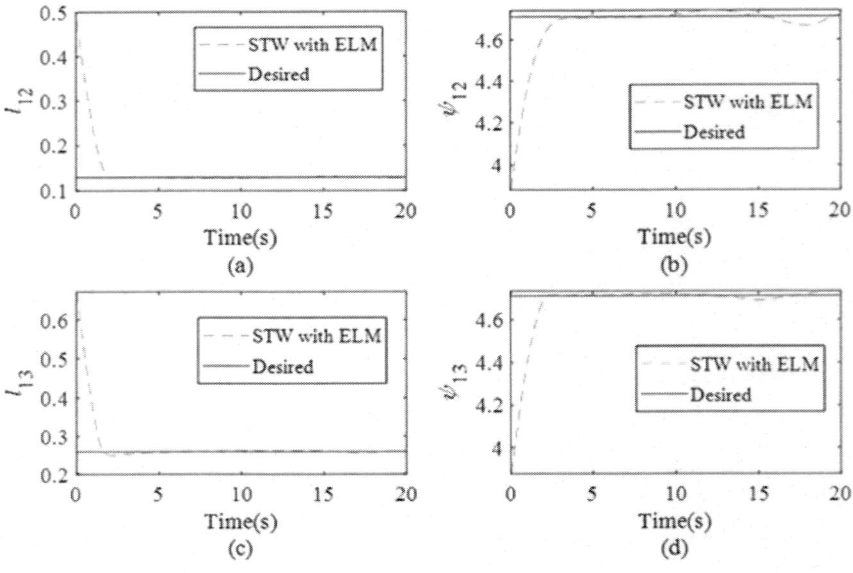

Figure 6.2. State variables by different methods. (a) l_{12} , (b) ψ_{12} , (c) l_{13} , (d) ψ_{13} .

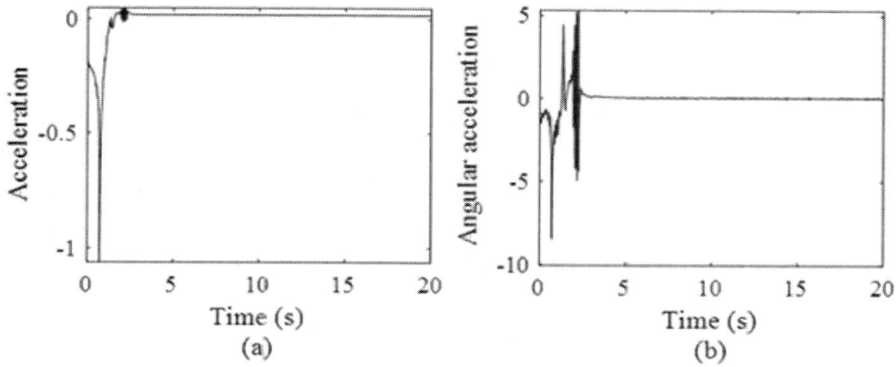

Figure 6.3. Control inputs from the follower 2. (a) Acceleration by STW with ELM and (b) angular acceleration by STW with ELM.

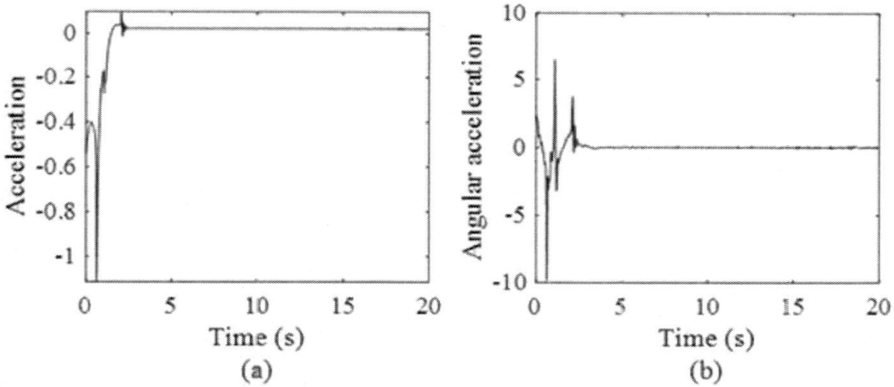

Figure 6.4. Control inputs from the follower 3. (a) Acceleration by STW with ELM and (b) angular acceleration by STW with ELM.

Figure 6.5 illustrates the sliding surfaces. The estimations of uncertainties and the estimation errors are illustrated in Figures 6.6 and 6.7. As proven in Theorem 1, the formation control system is asymptotically stable. From Figure 6.7, we know the errors are large at the outset but the errors are dramatically decreased while the weights of the ELM are adjusted by (6.36).

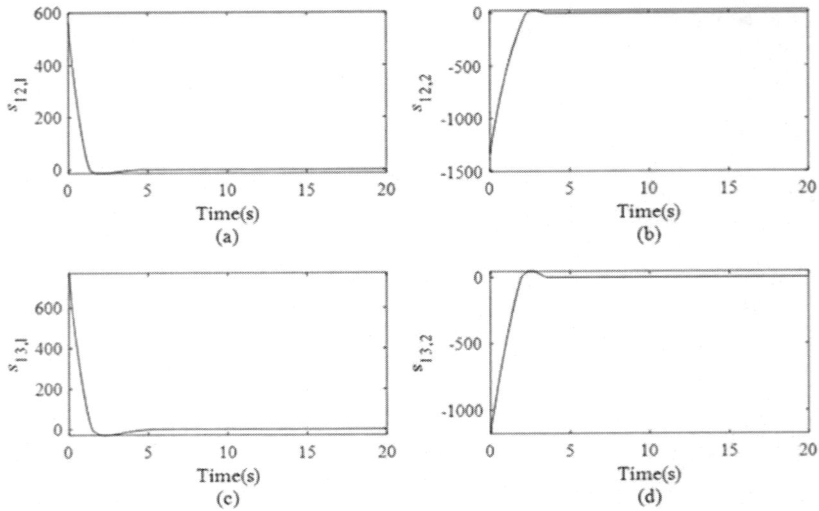

Figure 6.5. Sliding surfaces of the two followers. (a) $s_{12,1}$, (b) $s_{12,2}$, (c) $s_{13,1}$, (d) $s_{13,2}$.

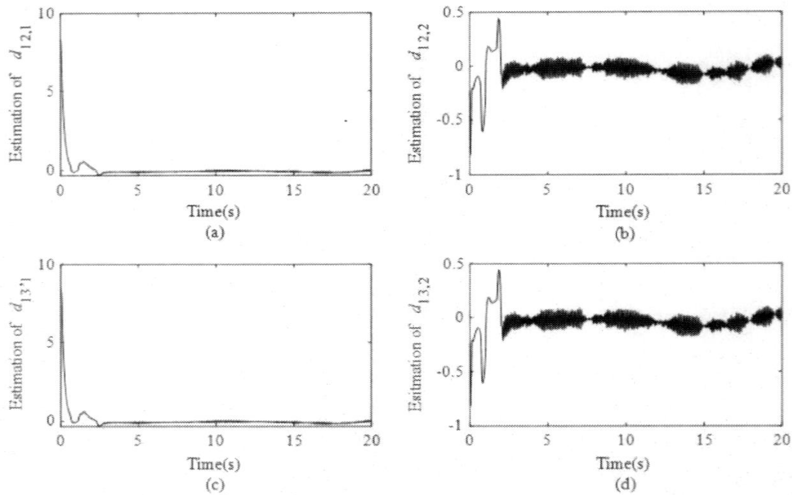

Figure 6.6. ELM outputs of the two followers. (a) Estimation of $d_{12,1}$, (b) Estimation of $d_{12,2}$, (c) Estimation of $d_{13,1}$, (d) Estimation of $d_{13,2}$.

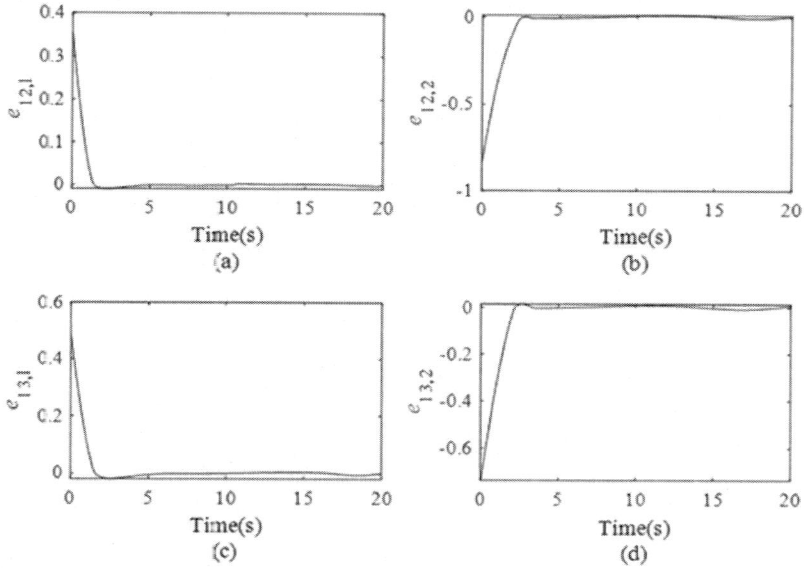

Figure 6.7. Estimation errors of the ELM for the two followers. (a) $e_{12,1}$, (b) $e_{12,2}$, (c) $e_{13,1}$, (d) $e_{13,2}$.

CONCLUSION

This paper has concentrated on the formation control of multi-robot systems. In order to accomplish the formation task despite the inevitable uncertainties and disturbances, the super-twisting sliding mode control method is adopted. For the sake of dealing with the overestimate of the control gains, the ELM is deduced. The weights of the ELM between the hidden and output layers can be adaptively adjusted. Theoretically, the integration of the super-twisting sliding mode control and the ELM for the formation maneuvers has the guaranteed stability in the sense of Lyapunov. In reality, the integrated method is applied to a multi-robot platform with three mobile robots. The numerical results illustrate that the integrated method has the good performance. This integrated method can be a solid support to deal with the formation maneuvers of uncertain multiple robots.

REFERENCES

[1] Wang, J. L., Wu, H. N. (2012). Leader-following formation control of multi-agent systems under fixed and switching topologies. *International Journal of Control*, 85(6), 695 - 705.

[2] Li, W., Chen, Z., Liu, Z. (2013). Leader-following formation control for second-order multiagent systems with time-varying delay and nonlinear dynamics. *Nonlinear Dynamics*, 72(4), 803 - 812.

[3] Xiao-Yuan, L., Na-Ni, H., Xin-Ping, G. (2010). Leader-following formation control of multi-agent networks based on distributed observers. *Chinese Physics B*, 19(10), 100202.

[4] Liang, X., Liu, Y. H., Wang, H., Chen, W., Xing, K., Liu, T. (2016). Leader-following formation tracking control of mobile robots without direct position measurements. *IEEE Transactions on Automatic Control*, 61(12), 4131 - 4137.

[5] Beard, R. W., Lawton, J., Hadaegh, F. Y. (2001). A coordination architecture for spacecraft formation control. *IEEE Transactions on control systems technology*, 9(6), 777 - 790.

[6] Gustavi, T., Hu, X. (2008). Observer-based leader-following formation control using onboard sensor information. *IEEE Transactions on Robotics*, 24(6), 1457 - 1462.

[7] Ren, W., Beard, R. (2002, August). Virtual structure based spacecraft formation control with formation feedback. In *Proceedings of AIAA Guidance, Navigation, and control conference and exhibit* (p. 4963).

[8] Wang, X., Ni, W., Wang, X. (2012). Leader-following formation of switching multirobot systems via internal model. *IEEE Transactions on Systems, Man, and Cybernetics, Part B (Cybernetics)*, 42(3), 817 - 826.

[9] Xiao, F., Wang, L., Chen, J., Gao, Y. (2009). Finite-time formation control for multi-agent systems. *Automatica*, 45(11), 2605 - 26.

[10] Han, T., Guan, Z. H., Chi, M., Hu, B., Li, T., Zhang, X. H. (2017). Multi-formation control of nonlinear leader-following multi-agent systems. *ISA transactions*, 69, 140 - 147.

[11] Luo, X., Han, N., Guan, X. (2011). Leader-following consensus protocols for formation control of multi-agent network. *Journal of Systems Engineering and Electronics*, 22(6), 991 - 997.

[12] Ren, W., Beard, R. W. (2004). Formation feedback control for multiple spacecraft via virtual structures. *IEE Proceedings-Control Theory and Applications*, 151(3), 357 - 368.

[13] Shi, Q., Li, T., Li, J., Chen, C. P., Xiao, Y., Shan, Q. (2019). Adaptive leader-following formation control with collision avoidance for a class of second-order nonlinear multi-agent systems. *Neurocomputing*, 350, 282 - 290.

[14] Wang, J., Wu, X., Xu, Z. (2008). Potential-based obstacle avoidance in formation control. *Journal of Control Theory and Applications*, 6(3), 311 - 316.

[15] Liu, Y., Jia, Y. (2012). An iterative learning approach to formation control of multi-agent systems. *Systems and Control Letters*, 61(1), 148 - 154.

[16] Wang, J., Xin, M. (2012). Integrated optimal formation control of multiple unmanned aerial vehicles. *IEEE Transactions on Control Systems Technology*, 21(5), 1731 - 1744.

[17] Lü, J., Chen, F., Chen, G. (2016). Nonsmooth leader-following formation control of nonidentical multi-agent systems with directed communication topologies. *Automatica*, 64, 112 - 120.

[18] Miao, Z., Liu, Y. H., Wang, Y., Yi, G., Fierro, R. (2018). Distributed estimation and control for leader-following formations of nonholonomic mobile robots. *IEEE Transactions on Automation Science and Engineering*, 15(4), 1946 - 1954.

[19] Chang, Y. H., Chang, C. W., Chen, C. L., Tao, C. W. (2011). Fuzzy sliding-mode formation control for multirobot systems: design and implementation. *IEEE Transactions on Systems, Man, and Cybernetics, Part B (Cybernetics)*, 42(2), 444 - 457.

[20] Patel, K., Mehta, A. (2020). Discrete-time sliding mode protocols for leader-following consensus of discrete multi-agent system with switching graph topology. *European Journal of Control*, 51, 65 - 75.

[21] Yu, D., Dong, L., Yan, H. (2019). Adaptive sliding mode control of multi-agent relay tracking systems with disturbances. *Journal of Control and Decision*, 1 - 16.

[22] Wang, J., Xu, Y., Xu, Y., Yang, D. (2019). Time-varying formation for high-order multi-agent systems with external disturbances by event-triggered integral sliding mode control. *Applied Mathematics and Computation*, 359, 333 - 343.

[23] Wang, J., Luo, X., Li, X., Zhu, M., Guan, X. (2019). Sliding Mode Formation Control of Nonlinear Multi-agent Systems with Local Lipschitz Continuous Dynamics. *Journal of Systems Science and Complexity*, 32(3), 759 - 777.

[24] Mishra, R. K., Sinha, A. (2019). Event-triggered sliding mode based consensus tracking in second order heterogeneous nonlinear multi-agent systems. *European Journal of Control,* 45, 30 - 44.

[25] Qian, D., Zhang, G., Chen, J., Wang, J., Wu, Z. (2019). Coordinated formation design of multi-robot systems via an adaptive-gain super-twisting sliding mode method. *Applied Sciences*, 9(20), 4315.

[26] Yaghoubi, Z., Talebi, H. A. (2019). Cluster consensus of general fractional-order nonlinear multi agent systems via adaptive sliding mode controller. *Archives of Control Sciences*, 29.

[27] Mirabdollahi, S. E. and Haeri, M. (2019). Multi-agent system finite-time consensus control in the presence of disturbance and input saturation by using of adaptive terminal sliding mode method. *Cogent Engineering*, 6(1), 1698689.

[28] Kamal, S., Sachan, A., Kumar, D. K., Singh, D. (2019). Robust finite time cooperative control of second order agents: A multi-input multi-output higher order super-twisting based approach. *ISA transactions*, 86, 1 - 8.

[29] Yaseen, Z. M., Sulaiman, S. O., Deo, R. C., Chau, K. W. (2019). An enhanced extreme learning machine model for river flow forecasting: State-of-the-art, practical applications in water resource engineering area and future research direction. *Journal of Hydrology*, 569, 387 - 408.

[30] Sattar, A. M., Ertuğrul, Ö. F., Gharabaghi, B., McBean, E. A., Cao, J. (2019). Extreme learning machine model for water network management. *Neural Computing and Applications*, 31(1), 157 - 169.

[31] Dai, H., Cao, J., Wang, T., Deng, M., Yang, Z. (2019). Multilayer one-class extreme learning machine. *Neural Networks*, 115, 11 - 22.

[32] Xiao, L. Y., Shao, W., Ding, X., Wang, B. Z. (2018). Dynamic adjustment kernel extreme learning machine for microwave component design. *IEEE Transactions on Microwave Theory and Techniques*, 66(10), 4452 - 4461.

[33] Yu, Y., Xu, M., Gu, J. (2019). Vision-based traffic accident detection using sparse spatio-temporal features and weighted extreme learning machine. *IET Intelligent Transport Systems*, 13(9), 1417 - 1428.

[34] Wong, K. I., Wong, P. K. (2018). Adaptive air-fuel ratio control of dual-injection engines under biofuel blends using extreme learning machine. *Energy conversion and management*, 165, 66 - 75.

[35] Wong, P. K., Wong, H. C., Vong, C. M., Xie, Z., Huang, S. (2016). Model predictive engine air-ratio control using online sequential extreme learning machine. *Neural Computing and Applications*, 27(1), 79 - 92.

In: A Closer Look at Formation Control ISBN: 978-1-53618-177-7
Editor: Dianwei Qian

Chapter 7

FORMATION DESIGN OF MULTIPLE AGENTS WITH UNCERTAINTIES BY INTEGRAL SLIDING MODE

Dianwei Qian and Chao Ma
School of Control and Computer Engineering,
North China Electric Power University, Beijing, China

ABSTRACT

This chapter investigates the formation-control problem of multiple agents. The formation control is based on leader-following approaches. The method of integral sliding mode control is adopted to achieve formation maneuvers of the agents on the concept of graph theory. Since the agents are subject to uncertainties, the uncertainties also challenge the formation-control design. Under a mild assumption that the uncertainties have an unknown bound, the technique of nonlinear disturbance observer is utilized to attract the issue. According to a given communication topology, formation stability conditions are investigated by the observer-based integral sliding mode formation control. In the sense of Lyapunov, not only the formation stability is guaranteed, but the desired formation of the agents is also realized. Finally, some simulation results are presented to show the feasibility and validity of the proposed control scheme through a multi-agent platform.

Keywords: multi-agent system, formation control, sliding mode control, uncertainty, nonlinear disturbance observer, stability

INTRODUCTION

Recently, multi-agent systems have been paid considerable attention [1−3]. The technology of multi-agent systems has been hailed as a novel paradigm for designing and implementing many real systems, such as distributed sensor networks, robots, satellite flying and aerial vehicles [4−6]. In reality, a multi-agent system has some advantages over a single agent, including but not limited to efficiency, reliability, extensibility, robustness and flexibility [7]. Increasing applications require multiple agents that have to work together. Such requirement rises up coordination of multiple agents [8, 9].

Among a variety of cooperative tasks, the consensus problem becomes attractive because it integrates both graph theory and control theory [10]. The consensus problem contains several typical control problems, i.e., cooperative control, formation control and flocking. See the report by Ren [11] for a complete review of recent philosophies in this area. As a branch, formation control works at controlling agents in a multi-agent system to form up and move in specified geometrical shapes [12, 13]. The background of formation control originates from the real world. For instants, the agents need to maintain some formations when they move at disaster sites, warehouses and hazardous areas [14].

One of coordinated control schemes for multi-agent formations is entitled ‘leader-following,’ where one agent is named leader and other agents are followers [15]. The sole leader is self-commanded and moves along a predefined trajectory. The followers take track of the leader. A cooperative task can be achieved by the interactions between these followers and the leader. The scheme has been successfully applied to the analysis and design of multiple agent systems [16]. However, the classic leader-following scheme is centralized, meaning that the control systems heavily depends on the leader and suffers from the 'single point of failure' problem [17, 18]. With the development of communication technology, it is desired to consider the communication topology in the scheme because the adaptability and practicability of multiple agents can be strengthened [19].

The leader-following scheme becomes emerging for multi-agent formations because its dynamics are experimentally modelled, but the internal formation stability can be theoretically guaranteed [20, 21]. Adopting such a scheme, various control methods have been developed for multi-agent formations, that is, robust control, dynamic output feedback method, adaptive fuzzy approach, multi-step predictive mechanism, iterative learning technique, neural network-based adaptive design, to name but a few. A systematic review on this topic is presented by Oh, Park and Ahn [22].

As a nonlinear feedback design tool, the sliding mode control (SMC) technique is popular because of its invariance, meaning that matched uncertainties in an SMC system can be suppressed by the invariance [23, 24]. Some SMC-based methods have been addressed to solve the formation-control problem of multi-agent systems, that is, first-order SMC, terminal SMC, backstepping SMC, e ct [25]. Previous contributions have verified the feasibility of the SMC methodology for multi-agent formations [26–27].

Compared to other SMC methods, the integral SMC design can guarantee an integral SMC system against matched uncertainties from the initial time, indicating that an entire response of the system is of invariance. Some successful applications of the integral SMC method have been reported in industries, for example, power systems, hypersonic vehicles as well as multi-agent systems [28–30].

Uncertainties exist everywhere. Each individual agent may suffer from uncertainties, i.e., external disturbances, unmodelled dynamics and parameter perturbations [31]. Originated from the uncertainties of individual agents, formation dynamics of multi-agent systems become uncertain [32]. In previous works about the SMC-based multi-agent formations, uncertainties are considered because they adversely affect the formation stability. Two solutions can be summarized from the works. One solution is to discuss the formation stability by means of graph theory. The other is to analyze the formation stability in light of Lyapunov's theorem. To guarantee the formation stability, both the solutions need to assume that the formation uncertainties are bounded. Unfortunately, the boundary of uncertainties is rather hard to exactly measure or know in advance. The lack of such a boundary may result in severe problems, i.e., decrease of the formation robustness, deterioration of the formation performance as far as deficiency of the formation stability [33, 34]. To obtain the important information, it is desired to adaptively approximate the formation uncertainties.

The technique of nonlinear disturbance observer (NDO) has been proven to be effective in handling uncertainties and improving robustness [35]. The applications of NDO have been investigated by some actual cases. Such a technique can be considered as an alternative to attack the issue of uncertainties [36]. So far, the academic problem of how to eliminate the adverse effects of uncertainties in multi-agent formations via NDO still remains unsolved and problematic.

This paper touches the academic problem. An NDO-based observer is employed to observe the uncertainties of agents. Based on the observer, an integral sliding mode controller is developed to achieve the formation maneuvers of agents. The formation stability is presented in the sense of Lyapunov. By a comparison, some numerical results illustrate that the presented control design is preferable.

MODELING

One Single Agent

A multi-agent system contains N identical agents. Displayed in Figure 7.1, the single agent under consideration is a differential wheeled robot in the horizon. The robot is round with a radius of r. Use index i to refer to the robot in Figure 7.1. The robot's movement is based on two separately driven wheels placed on either side of the robot's body, where $(x_{ci},\ y_{ci})$ are the center of this robot, $(x_{Li},\ y_{Li})$ are the center of the left wheel, $(x_{Ri},\ y_{Ri})$ are the center of the right wheel and are the center of the robot's body. Then, a vector $\mathbf{q}_i = [x_{hi}\ y_{hi}\ \theta_i]^T$ can be defined to describe the robot's posture, where (x_{hi}, y_{hi}) are the robot's head in the inertial frame and θ_i denotes the orientation angle of a mobile frame. Under the assumption that pure rolling and non-slipping motion, the robot is subjected to $[-\sin\theta_i\ \cos\theta_i\ 0]\cdot[x_{ci}\ y_{ci}\ \theta_i]=0$.

The Lagrangian equation of the agent can be obtained as $L_i = K_i - P_i$, where K_i denotes the kinetic energy of the agent and P_i means the potential energy. Concerning the agent, it can only move in the horizon so that its potential energy is kept unchanged, that is, $P_i \equiv 0$. Then, L_i can be written by

$$L_i = K_{bi} + K_{li} + K_{ri} \tag{7.1}$$

In (7.1), $K_{bi} = \frac{1}{2}m_b(\dot{x}_{ci}^2 + \dot{y}_{ci}^2) + \frac{1}{2}I_b\dot{\theta}_i^2$, $K_{li} = \frac{1}{2}m_w(\dot{x}_{li}^2 + \dot{y}_{li}^2) + \frac{1}{2}I_w\dot{\theta}_i^2$ and $K_{ri} = \frac{1}{2}m_w(\dot{x}_{ri}^2 + \dot{y}_{ri}^2) + \frac{1}{2}I_w\dot{\theta}_i^2$, where m_b and I_b are the mass and the moment of inertia of the robot's body, respectively; m_w and I_b are the mass and the moment of inertia of the robot's wheel, respectively.

The Lagrangian equations of motion can be formulated as (6.2) with respect to the vector $\mathbf{q}_i$.

$$\frac{\mathrm{d}}{\mathrm{d}t}\left(\frac{\partial L_i}{\partial \dot{\mathbf{q}}_i}\right) - \frac{\partial L_i}{\partial \mathbf{q}_i} = \mathbf{B}(\mathbf{q}_i)\tau_i \tag{7.2}$$

where $\tau_i = [\tau_{Li}\ \tau_{Ri}]^T$ is the torque vector applied to the wheels and $\mathbf{B}(\mathbf{q}_i)$ is a time-varying matrix. By the Lagrangian method, the dynamic model of the agent can be formulated by

$$\mathbf{M}(\mathbf{q}_i)\ddot{\mathbf{q}}_i + \mathbf{C}(\mathbf{q}_i, \dot{\mathbf{q}}_i)\dot{\mathbf{q}}_i = \mathbf{B}(\mathbf{q}_i)\tau_i \tag{7.3}$$

where the matrices $\mathbf{M}(\mathbf{q}_i)$, $\mathbf{C}(\mathbf{q}_i, \mathbf{q}_i)$ and $\mathbf{B}^T(\mathbf{q}_i)$ in order are determined by $\begin{bmatrix} m & 0 & mh\sin\theta_i \\ 0 & m & -mh\cos\theta_i \\ mh\sin\theta_i & -mh\cos\theta_i & I \end{bmatrix}$, $\begin{bmatrix} 0 & 0 & mh\dot{\theta}_i\cos\theta_i \\ 0 & 0 & -mh\dot{\theta}_i\sin\theta_i \\ 0 & 0 & 0 \end{bmatrix}$ and $\frac{1}{r}\begin{bmatrix} \cos\theta_i & \sin\theta_i & -1 \\ \cos\theta_i & \sin\theta_i & 1 \end{bmatrix}$.

In (7.3), $I = I_b + 2I_w + 2m_w l^2 + m_b h^2$, $m = m_b + 2m_w$, h is the length between the agent's center and the head and l is the distance between the agent's center and one wheel.

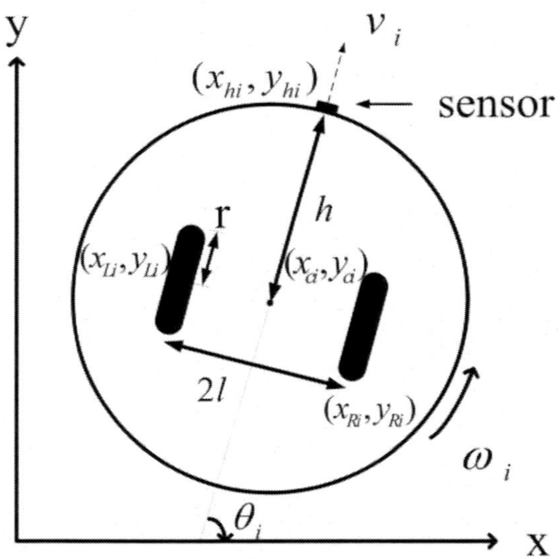

Figure 7.1. Schematic diagram of a single agent.

From Figure 7.1, two symbols of the agent are unexplained, that is, the linear velocity v_i and the rotation angular velocity ω_i. Define a vector $\xi_i = [v_i\ \omega_i]^T$. Then, $\dot{\mathbf{q}}_i = T(\mathbf{q}_i)\xi_i$ exists, here $\mathbf{T}^T(\mathbf{q}_i) = \begin{bmatrix} \cos\theta_i & \sin\theta_i & 0 \\ -h\sin\theta_i & h\cos\theta_i & 1 \end{bmatrix}$.

Substituting $\dot{\mathbf{q}}_i = T(\mathbf{q}_i)\xi_i$ into (7.3) yields

$$\tilde{\mathbf{M}}(\mathbf{q}_i)\dot{\xi}_i + \tilde{\mathbf{C}}(\mathbf{q}_i, \dot{\mathbf{q}}_i)\xi_i = \tilde{\mathbf{B}}(\mathbf{q}_i)\tau_i \tag{7.4}$$

Here $\tilde{\mathbf{M}}(\mathbf{q}_i)$, $\tilde{\mathbf{C}}(\mathbf{q}_i, \dot{\mathbf{q}}_i)$ and $\mathbf{B}(\mathbf{q}_i)$ in order are determined by $\mathbf{T}^T(\mathbf{q}_i)\mathbf{M}(\mathbf{q}_i)\mathbf{T}(\mathbf{q}_i)$, $\mathbf{T}^T(\mathbf{q}_i)[\mathbf{M}(\mathbf{q}_i)\dot{\mathbf{T}}(\mathbf{q}_i) + \mathbf{C}(\mathbf{q}_i, \dot{\mathbf{q}}_i)\mathbf{T}(\mathbf{q}_i)]$ and $\mathbf{T}^T(\mathbf{q}_i)\mathbf{B}(\mathbf{q}_i) = \frac{1}{r}\begin{bmatrix} 1 & 1 \\ -l & l \end{bmatrix}$.

In (7.4), $\tilde{\mathbf{C}}(\mathbf{q}_i, \dot{\mathbf{q}}_i) = \mathbf{0}_{2\times2}$ where $\mathbf{0}_{2\times2}$ is a 2×2 zero matrix. The fact indicates that $\tilde{\mathbf{M}}(\mathbf{q}_i)\dot{\xi}_i = \tilde{\mathbf{B}}(\mathbf{q}_i)\tau_i$ can be deduced from (7.4). Provided $\det[\mathbf{M}(\mathbf{q}_i)] \neq 0$, the assumption means $\tilde{\mathbf{M}}^{-1}(\mathbf{q}_i)$ is invertible such that the the equations of motion describing the behaviour of the agent can be formulated by

$$\dot{\xi}_i = \tilde{\mathbf{M}}^{-1}(\mathbf{q}_i)\tilde{\mathbf{B}}(\mathbf{q}_i)\tau_i \tag{7.5}$$

Reconsider $\dot{\mathbf{q}}_i = T(\mathbf{q}_i)\xi_i$ so that the equations of motion of the agent at its head has a form of

$$[\dot{x}_{hi}\ \dot{y}_{hi}]^T = \mathbf{H}[v_i\ \omega_i]^T \tag{7.6}$$

where $\mathbf{H} = \begin{bmatrix} \cos\theta_i & -h\sin\theta_i \\ \sin\theta_i & h\cos\theta \end{bmatrix}$.

Differentiating (7.6) with respect to time t yields

$$\begin{bmatrix} \ddot{x}_{hi} \\ \ddot{y}_{hi} \end{bmatrix} = \begin{bmatrix} \dot{v}_{xi} \\ \dot{v}_{yi} \end{bmatrix} = \mathbf{H}\tilde{\mathbf{M}}^{-1}(\mathbf{q}_i)\tilde{\mathbf{B}}(\mathbf{q}_i)\tau_i + \begin{bmatrix} \rho_{1i} \\ \rho_{2i} \end{bmatrix} \tag{7.7}$$

where $\rho_{1i} = -v_i\omega_i \sin\theta_i - h\omega_i^2 \cos\theta_i$ and $\rho_{2i} = v_i\omega_i \cos\theta_i - h\omega_i^2 \sin\theta_i$.

Define the control actions $[u_{xi}\ u_{yi}]^T = \mathbf{H}\tilde{\mathbf{M}}^{-1}(\mathbf{q}_i)\tilde{\mathbf{B}}(\mathbf{q}_i)\tau_i + [\rho_{1i}\ \rho_{2i}]^T$. (7.7) can have a form of

$$\begin{bmatrix} \ddot{x}_{hi} \\ \ddot{y}_{hi} \end{bmatrix} = \begin{bmatrix} \dot{v}_{xi} \\ \dot{v}_{yi} \end{bmatrix} = \begin{bmatrix} u_{xi} \\ u_{yi} \end{bmatrix}$$

Consider the agent's uncertainties, Finally, the agent can be described by

$$\begin{bmatrix} \dot{x}_{hi} \\ \dot{v}_{xi} \\ \dot{y}_{hi} \\ \dot{v}_{yi} \end{bmatrix} = \begin{bmatrix} v_{xi} \\ u_{xi} + \delta_{xi} \\ v_{yi} \\ u_{yi} + \delta_{yi} \end{bmatrix} \tag{7.8}$$

where δ_{xi} and δ_{yi} are the uncertain terms.

Assumption 1: $\|\delta_{xi}\| \le \delta_{xi}^*$ and $\|\delta_{yi}\| \le \delta_{yi}^*$, where $\delta_{xi}^* > 0$ and $\delta_{yi}^* > 0$ are constant but unknown.

Assumption 2: Both δ_{xi} and δ_{yi} are slowly time-varying, indicating $\dot{\delta}_{xi} \simeq 0$ and $\dot{\delta}_{yi} \simeq 0$.

Assumption 1 indicates the uncertainties are bounded by an unknown constant, which is mild. From Assumption 2, δ_{xi} and δ_{yi} seem almost constant. Considering the technical contents, Assumption 2 hints the designed observer could evaluate or calculate δ_{xi} and δ_{yi} much faster than the change rates of δ_{xi} and δ_{yi}. In this sense, δ_{xi} and δ_{yi} could be treated as almost constants by the observer.

Communication Topologies

Re-consider the multi-agent system with N agents. The communication topologies of such a system can be modelled via the theory of algebraic graph. Define a directed graph $\mathcal{G}=(\mathcal{V},\mathcal{E})$ composed of a vertex set $\mathcal{V}$ and an edge set $\mathcal{E}$, where $\mathcal{V}=\{v_1,v_2,\cdots,v_N\}$, $\mathcal{E}\subseteq\mathcal{V}\times\mathcal{V}$, the node v_i denotes the ith agent and $i=1,2,\cdots,N$. This paper investigates the directed graph in the multi-agent system. Consider the ith agent whose collection of neighbors is defined as $\mathcal{N}_i=\{v_j\in\mathcal{V}:(v_i,v_j)\in\mathcal{E}\}$. The ordered pair $(v_i,v_j)\in\mathcal{E}$ means that the jth agent can send information to the ith agent, but cannot send information vice versa.

The weighted adjacency matrix $\mathcal{A}$ of $\mathcal{G}$ has a form of

$$\mathcal{A}=\begin{bmatrix} a_{11} & a_{12} & \cdots & a_{1N} \\ a_{21} & a_{22} & \cdots & a_{2N} \\ \vdots & \vdots & \ddots & \vdots \\ a_{N1} & a_{N2} & \cdots & a_{NN} \end{bmatrix}\in\mathcal{R}^{N\times N} \tag{7.9}$$

where a_{ij} indicates the weight of the pair (v_i,v_j); $\forall\,(v_i,v_j)\in\mathcal{E},\exists\,a_{ij}=1$; $\forall\,(v_i,v_j)\notin\mathcal{E},\exists\,a_{ij}=0$; and $a_{ii}=0$.

The degree matrix of $\mathcal{G}$ is a diagonal matrix, determined by $\mathcal{D}=\mathrm{diag}\{d_1,d_2,\cdots,d_N\}\in\mathcal{R}^{N\times N}$. In the diagonal matrix, d_i is the in-degree of v_i, formulated by $d_i=\sum_{j=1}^{N}a_{ij}$ $(i=1,2,\cdots,N)$. Accordingly, the Laplacian matrix of $\mathcal{G}$ can be defined by $\mathcal{L}=\mathcal{D}-\mathcal{A}\in\mathcal{R}^{N\times N}$. Proven by Ren, $\mathcal{L}$ has at least one zero eigenvalue as well as all other eigenvalues are located at the open right-half plane if $\mathcal{G}$ is connected.

Assumption 3: $\mathcal{G}$ of the multi-agent system has a spanning tree, indicating that the zero eigenvalue of $\mathcal{L}$ is simple.

Assumption 4: In the multi-agent system, the sole leader cannot receive any information from the followers.

Consider Assumption 3. For the zero eigenvalue, an eigenvector of $\mathcal{L}$ is $\mathbf{1}_N$, that is, $\mathcal{L}\mathbf{1}_N = \mathbf{0}_N$ holds true, where $\mathbf{1}_N = [1, 1, \cdots, 1]^T \in \mathcal{R}^{N\times 1}$ and $\mathbf{0}_N = [0, 0, \cdots, 0]^T \in \mathcal{R}^{N\times 1}$. Further, $\text{rank}(\mathcal{L}) = N-1$ for the simple zero eigenvalue.

Without loss of generality, the *N*th agent in the multi-agent system is named leader and other $N-1$ agents are followers. Consider Assumption 4, that is, $a_{Ni} = 0\,(i = 1, 2, \cdots, N)$ and the Laplacian matrix of $\mathcal{G}$ can be written as

$$\mathcal{L} = \begin{bmatrix} d_1 & -a_{12} & \cdots & -a_{1(N-1)} & -a_{1N} \\ -a_{21} & d_2 & \cdots & -a_{2(n-1)} & -a_{2N} \\ \vdots & \vdots & \ddots & \vdots & \vdots \\ -a_{(N-1)1} & -a_{(N-2)2} & \cdots & d_{N-1} & -a_{(N-1)N} \\ 0 & 0 & 0 & 0 & 0 \end{bmatrix} \tag{7.10}$$

Further, the communication topologies among all the followers are described by a directed graph $\overline{\mathcal{G}}$. Apparently, $\overline{\mathcal{G}}$ is a subgraph of $\mathcal{G}$. The weighted adjacency matrix $\overline{\mathcal{A}} \in \mathcal{R}^{(N-1)\times(N-1)}$ of $\overline{\mathcal{G}}$ is defined by

$$\overline{\mathcal{A}} = \begin{bmatrix} a_{11} & a_{12} & \cdots & a_{1(N-1)} \\ a_{21} & a_{22} & \cdots & a_{2(N-1)} \\ \vdots & \vdots & \ddots & \vdots \\ a_{(N-1)1} & a_{(N-1)2} & \cdots & a_{(N-1)(N-1)} \end{bmatrix} \tag{7.11}$$

The degree matrix of $\overline{\mathcal{G}}$ is determined by $\overline{\mathcal{D}} = \text{diag}\{\overline{d}_1, \overline{d}_2, \cdots, \overline{d}_{N-1}\}$, where $\overline{d}_i = \sum_{j=1}^{N-1} a_{ij}\ (i = 1, 2, \cdots, N-1)$. Accordingly, the Laplacian matrix of $\overline{\mathcal{G}}$ can be defined as $\overline{\mathcal{D}} - \overline{\mathcal{A}}$, formulated by

$$\overline{\mathcal{L}} = \begin{bmatrix} \overline{d}_1 & -a_{12} & \cdots & -a_{1(N-1)} \\ -a_{21} & \overline{d}_2 & \cdots & -a_{2(N-2)} \\ \vdots & \vdots & \ddots & \vdots \\ -a_{(N-1)1} & -a_{(N-1)2} & \cdots & \overline{d}_{N-1} \end{bmatrix} \quad (7.12)$$

Similarly, assuming that the subgraph $\overline{\mathcal{G}}$ is itself a directed graph, $\overline{\mathcal{L}}\mathbf{1}_{N-1} = \mathbf{0}_{N-1}$ exists where $\mathbf{1}_{N-1} = [1, 1, \cdots, 1]^T \in \mathcal{R}^{(N-1)\times 1}$ and $\mathbf{0}_{N-1} = [0, 0, \cdots, 0]^T \in \mathcal{R}^{(N-1)\times 1}$. Moreover, define a matrix $\overline{\mathcal{B}} = \text{diag}\{\overline{b}_1, \overline{b}_2, \cdots, \overline{b}_{N-1}\}$, where $\overline{b}_i = a_{iN}$ $(i = 1, 2, \cdots, N-1)$. Apparently, there exists $\text{rank}(\overline{\mathcal{L}} + \overline{\mathcal{B}}) = \text{rank}(\mathcal{L}) = N-1$.

FORMATION DESIGN

Assumption 5: Considering the formation-control problem of the multi-agent system, the leader is assumed to be well controlled by a developed technology and the controller is known as well as the leader's its dynamics.

The *N*th agent has been named leader, its duty is to track a predefined trajectory and its control problem can be treated as the tracking-control problem of a single robot. Concerning Assumption 5, the leader can be treated as a nominal one in the formation-control problem, that is, $\delta_{xN} = \delta_{yN} = 0$. The other $N-1$ agents act as followers and they are equipped with the designed formation controllers to achieve formation maneuvers of the multi-agent system.

Concerning the *i*th follower agent $(i = 1, 2, \cdots, N-1)$, its equations of motion are decoupled in the *x*-axis and *y*-axis. Consequently, its formation design can be divided into two parts, that is, design for the *x*-axis and design for the *y*-axis. Here the formation design for the *x*-axis is taken into account at first. The *x*-axis subsystem with uncertainties can be written by

$$\begin{bmatrix} \dot{x}_{hi} \\ \dot{v}_{xi} \end{bmatrix} = \begin{bmatrix} v_{xi} \\ u_{xi} + \delta_{xi} \end{bmatrix} \tag{7.13}$$

Further, (7.13) can be re-written by the following state-space representation.

$$\dot{\mathbf{x}}_{xi} = \mathbb{A}_{xi}\mathbf{x}_{xi} + \mathbb{B}_{xi}u_{xi} + \mathbb{B}_{xi}\delta_{xi} \tag{7.14}$$

where $\mathbf{x}_{xi} = [x_{hi} \; v_{xi}]^T$, $\mathbb{A}_{xi} = \begin{bmatrix} 0 & 1 \\ 0 & 0 \end{bmatrix}$, $\mathbb{B}_{xi} = [0\,1]^T$ and u_{xi} is the control input.

NDO-Based Observer Design

Consider the x-axis subsystem (7.14) and design its NDO-based observer (7.15).

$$\begin{aligned} \dot{p}_{xi} &= -\mathbb{L}_{xi}^T\mathbb{B}_{xi}p_{xi} - \mathbb{L}_{xi}^T(\mathbb{B}_{xi}\mathbb{L}_{xi}^T\mathbf{x}_{xi} + \mathbb{A}_{xi}\mathbf{x}_{xi} + \mathbb{B}_{xi}u_{xi})\hat{\boldsymbol{\delta}}_{xi} \\ &= p_{xi} + \mathbb{L}_{xi}^T\mathbf{x}_{xi} \end{aligned} \tag{7.15}$$

In (7.15), p_{xi} is the internal state variable of the observer, $\hat{\delta}_{xi}$ is the approximation of δ_{xi} and the gain vector $\mathbb{L}_{xi} \in \mathcal{R}^{2\times 1}$ is designed such that the constant $\lambda_{xi} = \mathbb{L}_{xi}^T\mathbb{B}_{xi}$ is positive.

Define an estimation-error variable $e_{xi_d} = \delta_{xi} - \hat{\delta}_{xi}$, differentiate the error variable with respect to time t, substitute (7.15) into the derivative of e_{xi_d} and take Assumptions 1 and 2 into account. We can obtain (7.16).

$$\begin{aligned}
\dot{e}_{xi_d} &= \dot{\delta}_{xi} - \dot{\hat{\delta}}_{xi} \\
&= -\dot{p}_{xi} - \mathbb{L}_{xi}^T \dot{\mathbf{x}}_{xi} \\
&= \lambda_{xi} p_{xi} + \mathbb{L}_{xi}^T (\mathbb{B}_{xi} \mathbb{L}_{xi}^T \mathbf{x}_{xi} + \mathbb{A}_{xi} \mathbf{x}_{xi} + \mathbb{B}_{xi} u_{xi}) \\
&\quad - \mathbb{L}_{xi}^T (\mathbb{A}_{xi} \mathbf{x}_{xi} + \mathbb{B}_{xi} u_{xi} + \mathbb{B}_{xi} \delta_{xi}) \\
&= \lambda_{xi} (\hat{\delta}_{xi} - \mathbb{L}_{xi}^T \mathbf{x}_{xi}) + \mathbb{L}_{xi}^T (\mathbb{B}_{xi} \mathbb{L}_{xi}^T \mathbf{x}_{xi} + \mathbb{A} \mathbf{x}_{xi} + \mathbb{B} u_{xi}) \\
&\quad - \mathbb{L}_{xi}^T (\mathbb{A}_{xi} \mathbf{x}_{xi} + \mathbb{B}_{xi} u_{xi} + \mathbb{B}_{xi} \delta_{ik}) \\
&= \lambda_{xi} (\hat{\delta}_{xi} - \delta_{xi}) \\
&= -\lambda_{xi} e_{xi_d}
\end{aligned} \tag{7.16}$$

The solution of (7.16) is $e_{xi_d} = \exp(-\lambda_{xi} t) e_{xi_d}(0)$, where $e_{xi_d}(0)$ is the initial condition at $t = 0$. Owing to $\lambda_{xi} > 0$, this fact indicates that the estimation-error variable e_{xi_d} is exponentially convergent to 0 as $t \to \infty$.

Integral SMC-Based Controller Design

To coordinate the leader-following-based multi-agent system, a tracking-error variable of the *i*th follower agent is defined by

$$\begin{aligned}
e_{xi} &= \sum_{j=1}^{N-1} a_{ij} [(x_{hi} - x_{hj} - d_{ij}^x) + \rho_{xi}(v_{xi} - v_{xj})] \\
&\quad + \bar{b}_i (x_{hi} - x_{hN} - d_{iN}^x) + \bar{b}_i \rho_{xi} (v_{xi} - v_{xN})
\end{aligned} \tag{7.17}$$

where $\rho_{xi} > 0$ is a pre-defined constant, d_{ij}^x is the pre-defined relative position between the *i*th follower and the *j*th follower and d_{iN}^x is the pre-defined relative position between the *i*th follower and the leader.

Differentiate e_{xi} in (7.17) with respect to time *t*, substitute the *x*-axis subsystem (7.13) into the derivative of e_{xi} and consider Assumption 5 yields

$$\begin{aligned}\dot{e}_{xi} &= \sum_{j=1}^{N-1} a_{ij}[(v_{xi} - v_{xj}) + \rho_{xi}(u_{xi} - u_{xj}) + \rho_{xi}(\delta_{xi} - \delta_{xj})] \\ &+ \bar{b}_i[(v_{xi} - v_{xN}) + \rho_{xi}(u_{xi} - u_{xN}) + \rho_{xi}\delta_{xi}]\end{aligned} \tag{7.18}$$

Concerning the x-axis subsystem of the ith agent, an integral sliding-mode surface with the NDO-based observer output is defined by

$$s_{xi} = e_{xi} + c_{xi}\int e_{xi}\mathrm{d}t + \hat{\delta}_{xi} \tag{7.19}$$

here $c_{xi} > 0$ is constant.

The derivative of the sliding-mode surface variable with respect to time t can be written by

$$\dot{s}_{xi} = \dot{e}_{xi} + c_{xi}e_{xi} + \dot{\hat{\delta}}_{xi} \tag{7.20}$$

Substituting (7.18) and (7.19) into (7.20) gives

$$\begin{aligned}\dot{s}_{xi} &= \sum_{j=1}^{N-1} a_{ij}[(v_{xi} - v_{xj}) + \rho_{xi}(u_{xi} - u_{xj}) + \rho_{xi}(\delta_{xi} - \delta_{xj})] \\ &+ \bar{b}_i[(v_{xi} - v_{xN}) + \rho_{xi}(u_{xi} - u_{xN}) + \rho_{xi}\delta_{xi}] \\ &+ c_{xi}\sum_{j=1}^{N-1} a_{ij}[(x_{hi} - x_{hj} - d_{ij}^x) + \rho_{xi}(v_{xi} - v_{xj})] \\ &+ \bar{b}_i c_{xi}(x_{hi} - x_{hN} - d_{iN}^x) + \bar{b}_i c_{xi}\rho_{xi}(v_{xi} - v_{xN}) + \dot{\hat{\delta}}_{xi}\end{aligned} \tag{7.21}$$

Design the following x-axis formation-control law for the ith follower agent

$$\begin{aligned}u_{xi} &= -\frac{c_{xi}\rho_{xi} + 1}{\rho_{xi}(\bar{d}_i + \bar{b}_i)}\sum_{j=1}^{N-1} a_{ij}(v_{xi} - v_{xj}) - \frac{c_{xi}\rho_{xi} + 1}{\rho_{xi}(\bar{d}_i + \bar{b}_i)}\bar{b}_i(v_{xi} - v_{xN}) \\ &- \frac{\bar{b}_i}{\bar{d}_i + \bar{b}_i}\hat{\delta}_{xi} - \frac{1}{\bar{d}_i + \bar{b}_i}\sum_{j=1}^{N-1} a_{ij}(\hat{\delta}_{xi} - \hat{\delta}_{xj}) - \frac{c_{xi}}{\rho_{xi}(\bar{d}_i + \bar{b}_i)}\sum_{j=1}^{N-1} a_{ij}(x_{hi} - x_{hj} - d_{ij}^x) \\ &- \frac{\bar{b}_i c_{xi}}{\rho_{xi}(\bar{d}_i + \bar{b}_i)}(x_{hi} - x_{hN} - d_{iN}^x) + \frac{1}{\bar{d}_i + \bar{b}_i}\sum_{j=1}^{N-1} a_{ij}u_{xj} - \frac{\kappa_{xi}}{\rho_{xi}(\bar{d}_i + \bar{b}_i)}\mathrm{sgn}(s_{xi})\end{aligned} \tag{7.22}$$

where $\kappa_{xi} > 0$ is predefined and $\text{sgn}(\cdot)$ is the sign function.

The equation $\dot{\hat{\delta}}_{xi} = -\lambda_{xi}(\hat{\delta}_{xi} - \delta_{xi})$ can be drawn from (7.16). Replace $\dot{\hat{\delta}}_{xi}$ in (7.21) by the equation and replace u_{xi} in (7.21) by (7.22). Then, $\dot{s}_{xi}$ can be re-arranged by

$$\dot{s}_{xi} = \rho_{xi}(\bar{d}_i + \bar{b}_i)e_{xi_d} - \rho_{xi}\sum_{j=1}^{N-1} a_{ij}e_{xj_d} - \kappa_{xi}\,\text{sgn}(s_{xi}) - \lambda_{xi}(\hat{\delta}_{xi} - \delta_{xi}) \tag{7.23}$$

System Structure

The system structure is presented in Figure 7.2. Concerning the ith follower agent, its state-variable vector of the x-axis subsystem $\mathbf{x}_{xi}$, composed of x_{hi} and v_{xi}, is located at the feedback channel. From Figure 7.2, it feeds on the NDO-based observer to estimate the approximation $\hat{\delta}_{xi}$. Furthermore, the tracking-error variable e_{xi} and its integral construct the integral sliding-mode surface variable s_{xi}. The designed control scheme employs s_{xi}, $\mathbf{x}_{xi}$ and $\hat{\delta}_{xi}$ to generate the final control input u_{xi}. Finally, u_{xi} is applied to the ith follower agent to achieve formation maneuvers of its x-axis subsystem.

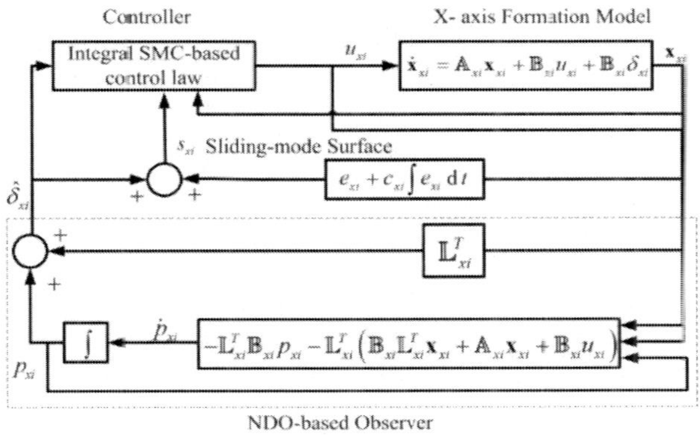

Figure 7.2. Structure of the control scheme.

Stability Analysis

In order to consider the formation stability of the x-axis subsystem, the following assumption about the estimation-error variable is considered.

Assumption 6: $|e_{xi_d}| < e^*_{xi_d}$, where $e^*_{xi_d} > 0$ is constant but unknown.

Theorem 1: Concern the ith follower agent, consider its x-axis subsystem dynamic (7.13), take Assumptions 1--6 into account, design the NDO-based observer (7.15), define the sliding-mode surface (7.19) and utilize the integral SMC-based control law (7.22). The closed-loop control system of the x-axis subsystem structured by Figure 7.2 is asymptotically stable if $\kappa_{xi} > (\lambda_{xi} + \rho_{xi}\bar{b}_i)e^*_{xi_d}$.

Proof: Select a Lyapunov candidate function $V = \frac{1}{2}s_{xi}^2$. Differentiate V with respect to time t. The derivative of V can be written by $\dot{V} = s_{xi}\dot{s}_{xi}$. Replace $\dot{s}_{xi}$ by (7.24). The derivative of V has the form of

$$\begin{aligned}
\dot{V} &= s_{xi}[\rho_{xi}(\bar{d}_i + \bar{b}_i)e_{xi_d} - \rho_{xi}\sum_{j=1}^{N-1} a_{ij}e_{xj_d} - \kappa_{xi}\,\mathrm{sgn}(s_{xi}) - \lambda_{xi}(\hat{\delta}_{xi} - \delta_{xi})] \\
&= s_{xi}[\rho_{xi}(\bar{d}_i + \bar{b}_i)e_{xi_d} - \rho_{xi}\sum_{j=1}^{N-1} a_{ij}e_{xi_d} - \kappa_{xi}\,\mathrm{sgn}(s_{xi}) + \lambda_{xi}e_{xi_d}] \\
&= -\kappa_{xi}|s_{xi}| + [\lambda_{xi}e_{xi_d} + \rho_{xi}(\bar{d}_i + \bar{b}_i)e_{xi_d} - \rho_{xi}\sum_{j=1}^{N-1} a_{ij}e_{xi_d}]s_{xi} \\
&\le -\kappa_{xi}|s_{xi}| + [\lambda_{xi}e^*_{xi_d} + \rho_{xi}(\bar{d}_i + \bar{b}_i)e^*_{xi_d} - \rho_{xi}\bar{d}_i e^*_{xi_d}]s_{xi} \\
&= -\kappa_{xi}|s_{xi}| + (\lambda_{xi} + \rho_{xi}\bar{b}_i)e^*_{xi_d}s_{xi} \\
&\le [-\kappa_{xi} + (\lambda_{xi} + \rho_{xi}\bar{b}_i)e^*_{xi_d}]|s_{xi}|
\end{aligned} \tag{7.24}$$

Pick up $\kappa_{xi} > (\lambda_{xi} + \rho_{xi}\bar{b}_i)e^*_{xi_d}$ so that $\dot{V} < 0$ exists. Concerning $V \ge 0$, the closed-loop control system of the x-axis subsystem is asymptotically stable in the sense of Lyapunov.

Furthermore, the closed-loop stability of the x-axis subsystem can be extended to the multi-agent system. Concerning the ith follower agent, define $\mathbf{z}_i = [x_{hi}\ y_{hi}]^T \in \mathcal{R}^{2\times1}$ so that the augmented vector of $\mathbf{z}_i$ in the system can be written by $\mathbf{z} = [\mathbf{z}_1^T\ \mathbf{z}_2^T \cdots \mathbf{z}_{N-1}^T]^T \in \mathcal{R}^{2(N-1)\times1}$.

Similarly, define the following vectors $\mathbf{v}_i = [v_{xi}\ v_{yi}]^T$, $\mathbf{u}_i = [u_{xi}\ u_{yi}]^T$, $\Delta_i = [\delta_{xi}\ \delta_{yi}]^T$, $\hat{\Delta}_i = [\hat{\delta}_{xi}\ \hat{\delta}_{yi}]^T$, $\mathbf{e}_i = [e_{xi}\ e_{yi}]^T$, $\mathbf{e}_{i_d} = [e_{xi_d}\ e_{yi_d}]^T$, $\mathbf{d}_{ij} = [d_{ij}^x\ d_{ij}^y]^T$, $\mathbf{d}_{iN} = [d_{iN}^x\ d_{iN}^y]^T$ and $\mathbf{s}_i = [s_{xi}\ s_{yi}]^T$. Here $\mathbf{v}_i$, $\mathbf{u}_i$, Δ_i, $\hat{\Delta}_i$, $\mathbf{e}_i$, $\mathbf{e}_{i_d}$, $\mathbf{d}_{ij}$, $\mathbf{d}_{iN}$ and $\mathbf{s}_i \in \mathcal{R}^{2\times1}$. Correspondingly, their augmented vectors are determined by $\mathbf{v} = [\mathbf{v}_1^T\ \mathbf{v}_2^T \cdots \mathbf{v}_{N-1}^T]^T$, $\mathbf{u} = [\mathbf{u}_1^T\ \mathbf{u}_2^T \cdots \mathbf{u}_{N-1}^T]^T$, $\Delta = [\Delta_1^T\ \Delta_2^T \cdots \Delta_{N-1}^T]^T$, $\hat{\Delta} = [\hat{\Delta}_1^T\ \hat{\Delta}_2^T \cdots \hat{\Delta}_{N-1}^T]^T$, $\mathbf{e} = [\mathbf{e}_1^T\ \mathbf{e}_2^T \cdots \mathbf{e}_{N-1}^T]^T$, $\mathbf{e}_d = [\mathbf{e}_{1_d}^T\ \mathbf{e}_{2_d}^T \cdots \mathbf{e}_{(N-1)_d}^T]^T$, $\mathbf{d}_i = [\mathbf{d}_{i1}^T\ \mathbf{d}_{i2}^T \cdots \mathbf{d}_{i(N-1)}^T]^T$, $\mathbf{d}_N = [\mathbf{d}_{1N}^T\ \mathbf{d}_{2N}^T \cdots \mathbf{d}_{(N-1)N}^T]^T$ and $\mathbf{s} = [\mathbf{s}_1^T\ \mathbf{s}_2^T \cdots \mathbf{s}_{N-1}^T]^T$. Here $\mathbf{v}$, $\mathbf{u}$, Δ, $\hat{\Delta}$, $\mathbf{e}$, $\mathbf{e}_d$, $\mathbf{d}_i$, $\mathbf{d}_N$ and $\mathbf{s} \in \mathcal{R}^{2(N-1)\times1}$.

Define $\Upsilon = \mathrm{diag}\{\rho_{x1}, \rho_{y1}, \cdots, \rho_{x(N-1)}, \rho_{y(N-1)}\}$, $\mathbf{c} = \mathrm{diag}\{c_{x1}, c_{y1}, \cdots, c_{x(N-1)}, c_{y(N-1)}\}$, $\Lambda = \mathrm{diag}\{\lambda_{x1}, \lambda_{y1}, \ldots, \lambda_{x(n-1)}, \lambda_{y(n-1)}\}$ and $\kappa = \mathrm{diag}\{\kappa_{x1}, \kappa_{y1}, \cdots, \kappa_{x(N-1)}, \kappa_{y(N-1)}\}$. Here Υ, $\mathbf{c}$, Λ and $\kappa \in \mathcal{R}^{2(N-1)\times2(N-1)}$.

From the above vectors and matrices, the augmented tracking-error vector $\mathbf{e}$ can be written by

$$\begin{aligned}\mathbf{e} &= [(\overline{\mathcal{L}} + \overline{\mathcal{B}}) \otimes \mathbf{I}_2](\mathbf{z} - \mathbf{d}_i) + \Upsilon[(\overline{\mathcal{L}} + \overline{\mathcal{B}}) \otimes \mathbf{I}_2]\mathbf{v} \\ &\quad -(\overline{\mathcal{B}} \otimes \mathbf{I}_2)(\mathbf{1}_{N-1} \otimes \mathbf{z}_N - \mathbf{d}_N) - \Upsilon(\overline{\mathcal{B}}\mathbf{1}_{N-1} \otimes \mathbf{I}_2)\mathbf{v}_n\end{aligned} \tag{7.25}$$

where $\mathbf{I}_2$ is a 2×2 identity matrix and $\otimes$ means the Kronecker product. Both of $\mathbf{z}_N$ and $\mathbf{v}_N$ are concerned to the leader agent, determined by $\mathbf{z}_N = [x_{hN}\ y_{hN}]^T$ and $\mathbf{v}_N = [v_{xN}\ v_{yN}]^T$.

Differentiating $\mathbf{e}$ in (7.25) with respect to time t gives

$$\begin{aligned}\dot{\mathbf{e}} &= [(\overline{\mathcal{L}} + \overline{\mathcal{B}}) \otimes \mathbf{I}_2]\mathbf{v} + \Upsilon[(\overline{\mathcal{L}} + \overline{\mathcal{B}}) \otimes \mathbf{I}_2]\mathbf{u} \\ &\quad +\Upsilon[(\overline{\mathcal{L}} + \overline{\mathcal{B}}) \otimes \mathbf{I}_2]\Delta - (\overline{\mathcal{B}}\mathbf{1}_{N-1} \otimes \mathbf{I}_2)\mathbf{v}_N - \Upsilon(\overline{\mathcal{B}}\mathbf{1}_{N-1} \otimes \mathbf{I}_2)\mathbf{u}_N\end{aligned} \tag{7.26}$$

where $\mathbf{u}_N$ is formulated by $\mathbf{u}_N = [u_{xN}\ u_{yN}]^T$.

The augmented integral sliding-mode surface vector is formulated by

$$\mathbf{s} = \mathbf{e} + \mathbf{c}\int \mathbf{e}dt + \hat{\Delta} \tag{7.27}$$

Differentiating $\mathbf{s}$ in (7.27) with respect to time t yields

$$\begin{aligned}\dot{\mathbf{s}} &= \dot{\mathbf{e}} + \mathbf{ce} + \dot{\hat{\Delta}} \\ &= [(\overline{\mathcal{L}} + \overline{\mathcal{B}}) \otimes \mathbf{I}_2]\mathbf{v} + \mathrm{Y}[(\overline{\mathcal{L}} + \overline{\mathcal{B}}) \otimes \mathbf{I}_2]\mathbf{u} \\ &\quad + \mathrm{Y}[(\overline{\mathcal{L}} + \overline{\mathcal{B}}) \otimes \mathbf{I}_2]\Delta - (\overline{\mathcal{B}}\mathbf{1}_{N-1} \otimes \mathbf{I}_2)\mathbf{v}_N - \mathrm{Y}(\overline{\mathcal{B}}\mathbf{1}_{N-1} \otimes \mathbf{I}_2)\mathbf{u}_N \\ &\quad + \mathbf{c}[(\overline{\mathcal{L}} + \overline{\mathcal{B}}) \otimes \mathbf{I}_2](\mathbf{z} - \mathbf{d}_i) + \mathbf{c}\mathrm{Y}[(\overline{\mathcal{L}} + \overline{\mathcal{B}}) \otimes \mathbf{I}_2]\mathbf{v} \\ &\quad - \mathbf{c}\mathrm{Y}(\overline{\mathcal{B}}\mathbf{1}_{N-1} \otimes \mathbf{I}_2)\mathbf{v}_N - \mathbf{c}(\overline{\mathcal{B}} \otimes \mathbf{I}_2)(\mathbf{1}_{N-1} \otimes \mathbf{z}_N - \mathbf{d}_N) + \dot{\hat{\Delta}}\end{aligned} \tag{7.28}$$

Design the control law $\mathbf{u}$ of the multi-agent system as

$$\begin{aligned}\mathbf{u} &= -\mathrm{Y}^{-1}[(\overline{\mathcal{L}} + \overline{\mathcal{B}}) \otimes \mathbf{I}_2]^{-1}\{\mathbf{c}[(\overline{\mathcal{L}} + \overline{\mathcal{B}}) \otimes \mathbf{I}_2](\mathbf{z} - \mathbf{d}) \\ &\quad + (\mathbf{I}_{2(N-1)} + \mathbf{c}\mathrm{Y})[(\overline{\mathcal{L}} + \overline{\mathcal{B}}) \otimes \mathbf{I}_2]\mathbf{v} \\ &\quad - (\mathbf{I}_{2(N-1)} + \mathbf{c}\mathrm{Y})(\overline{\mathcal{B}}\mathbf{1}_{N-1} \otimes \mathbf{I}_2)\mathbf{v}_N - \mathbf{c}(\overline{\mathcal{B}} \otimes \mathbf{I}_2)(\mathbf{1}_{N-1} \otimes \mathbf{z}_N - \mathbf{d}_N) \\ &\quad - \mathrm{Y}(\overline{\mathcal{B}}\mathbf{1}_{N-1} \otimes \mathbf{I}_2)\mathbf{u}_N + \mathrm{Y}[(\overline{\mathcal{L}} + \overline{\mathcal{B}}) \otimes \mathbf{I}_2]\hat{\Delta} + \kappa\,\mathrm{sgn}(\mathbf{s})\}\end{aligned} \tag{7.29}$$

In (7.29), $\mathbf{I}_{2(N-1)}$ is a $2(N-1) \times 2(N-1)$ identity matrix. Replacing the control law $\mathbf{u}$ in (7.28) by (7.29) gives

$$\dot{\mathbf{s}} = \mathrm{Y}[(\overline{\mathcal{L}} + \overline{\mathcal{B}}) \otimes \mathbf{I}_2](\Delta - \hat{\Delta}) - \kappa\,\mathrm{sgn}(\mathbf{s}) + \dot{\hat{\Delta}} \tag{7.30}$$

Theorem 2: Concern the multi-agent system, suppose that its communication graph has a directed spanning tree. The stability of the leader-following formation control is guaranteed if the controller parameters of each follower agent are designed by Theorem 1.

Proof: Define a Lyapunov candidate function $V'(t) = \|\mathbf{s}\|_2$ here $\|\cdot\|_2$ means 2-norm. Differentiate $V'(t)$ with respect to time t. The derivative of V' can be written by $\dot{V}'(t) = \frac{\mathbf{s}^T\dot{\mathbf{s}}}{\|\mathbf{s}\|_2}$. Replacing $\dot{\mathbf{s}}$ in the derivative of V' by (7.30) yields

$$\begin{aligned}
\dot{V}' &= \frac{\mathbf{s}^T}{\|\mathbf{s}\|_2}\{\Upsilon[(\overline{\mathcal{L}}+\overline{\mathcal{B}})\otimes\mathbf{I}_2](\Delta-\hat{\Delta})-\kappa\,\mathrm{sgn}(\mathbf{s})+\dot{\hat{\Delta}}\} \\
&\simeq \frac{\mathbf{s}^T}{\|\mathbf{s}\|_2}\{\Upsilon[(\overline{\mathcal{L}}+\overline{\mathcal{B}})\otimes\mathbf{I}_2]\mathbf{e}_d-\kappa\,\mathrm{sgn}(\mathbf{s})+\Lambda\mathbf{e}_d\} \\
&\le -\|\kappa\|_1\frac{\mathbf{s}^T\,\mathrm{sgn}(\mathbf{s})}{\|\mathbf{s}\|_2}+\|\Lambda\|_1\frac{\mathbf{s}^T\mathbf{e}_d}{\|\mathbf{s}\|_2}+\|\Upsilon\|_1\frac{\mathbf{s}^T[(\overline{\mathcal{L}}+\overline{\mathcal{B}})\otimes\mathbf{I}_2]\mathbf{e}_d}{\|\mathbf{s}\|_2} \\
&\le -\|\kappa\|_1\frac{\|\mathbf{s}\|_1}{\|\mathbf{s}\|_2}+\|\Lambda\|_1\frac{\mathbf{s}^T\mathbf{1}_{2(N-1)}}{\|\mathbf{s}\|_2}e_d^*+\|\Upsilon\|_1\frac{\mathbf{s}^T[(\overline{\mathcal{L}}+\overline{\mathcal{B}})\otimes\mathbf{I}_2]\mathbf{1}_{2(N-1)}}{\|\mathbf{s}\|_2}e_d^*
\end{aligned} \tag{7.31}$$

In (7.31), $e_d^* = \|\mathbf{e}_d\|_\infty$ where $\|\cdot\|_\infty$ means ∞-norm. Note $a_{ii} = 0\,(i = 1,\cdots,N-1)$ in (7.11) so that $[(\overline{\mathcal{L}}+\overline{\mathcal{B}})\otimes\mathbf{I}_2]\mathbf{1}_{2(N-1)} = [\overline{b}_1\,\overline{b}_1\cdots\overline{b}_{N-1}\,\overline{b}_{N-1}]^T \in \mathcal{R}^{2(N-1)}$. Let $b^* = \max\{\overline{b}_1,\cdots,\overline{b}_{N-1}\} = \|\mathcal{B}\|_\infty$. (7.31) can be re-arranged by

$$\begin{aligned}
\dot{V}' &\le -\|\kappa\|_1\frac{\|\mathbf{s}\|_1}{\|\mathbf{s}\|_2}+\|\Lambda\|_1\frac{\mathbf{s}^T\mathbf{1}_{2(N-1)}}{\|\mathbf{s}\|_2}e_d^*+\|\Upsilon\|_1\frac{\mathbf{s}^T\mathbf{1}_{2(N-1)}}{\|\mathbf{s}\|_2}b^*e_d^* \\
&\le -\|\kappa\|_1\frac{\|\mathbf{s}\|_1}{\|\mathbf{s}\|_2}+\|\Lambda\|_1\frac{\|\mathbf{s}\|_1}{\|\mathbf{s}\|_2}e_d^*+\|\Upsilon\|_1\frac{\|\mathbf{s}\|_1}{\|\mathbf{s}\|_2}b^*e_d^*
\end{aligned} \tag{7.32}$$

If the controller parameters of each follower agent are selected by Theorem 1, $\dot{V}' < 0$ can be deduced from (7.32). Considering $V' \ge 0$, the formation control of the multi-agent system is asymptotically stable in the sense of Lyapunov.

Take Assumption 6 into account. The tracking-error variable $e^*_{xi_d}$ is constant but unknown, meaning that it is hard to determine κ_{ik} in Theorem 1 as well as κ in Theorem 2. To guarantee the formation stability, a conservative value of $e^*_{xi_d}$ should be assigned. On this aspect, there seem no benefits earned from such an NDO-based integral SMC scheme. However, e_{xi_d} originated from the scheme is exponentially convergent as proven, indicating that a small value of $e^*_{xi_d}$ could be chosen. According to Theorem 1, such a kind of formation design could contribute to the decrease of the chattering phenomenon as well as the improvement of the formation performance.

SIMULATION RESULTS

To demonstrate the performance of the presented control scheme, this section implements some simulations on a multi-robot platform and discusses the results. The platform is composed of four mobile wheeled robots. The robots are identical (See Figuer 7.1), where three follower agents in order are numbered by 1, 2 and 3 and the sole leader agent is numbered by 4. The physical parameters of these agents are picked up, listed by $l = 0.0265\,\text{m}$, $h = 0.04\,\text{m}$, $r = 0.02\,\text{m}$, $m_b = 0.018\,\text{kg}$, $m_w = 0.007\,\text{kg}$, $I_b = 1.44\times 10^{-4}\,\text{kg}\cdot\text{m}^2$ and $I_w = 1.44\times 10^{-6}\,\text{kg}\cdot\text{m}^2$. The communication topology of this multi-agent system is illustrated in Figure 7.3.

From Figure 7.3, the communication graph $\mathcal{G}$ forms a standard spanning tree, where the adjacency and Laplacian matrices are formulated by

$$\mathcal{A} = \begin{bmatrix} 0 & 1 & 0 & 1 \\ 1 & 0 & 0 & 0 \\ 0 & 0 & 0 & 1 \\ 0 & 0 & 0 & 0 \end{bmatrix} \text{ and } \mathcal{L} = \begin{bmatrix} 2 & -1 & 0 & -1 \\ -1 & 1 & 0 & 0 \\ 0 & 0 & 1 & -1 \\ 0 & 0 & 0 & 0 \end{bmatrix}$$

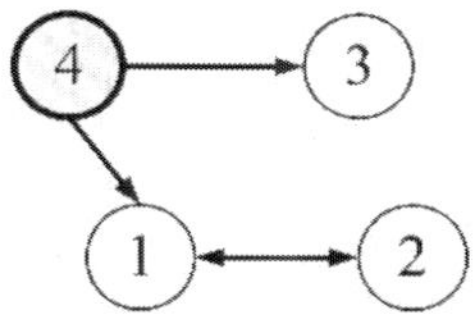

Figure 7.3. Communication topology of the multi-agent platform.

Further, the communication subgraph $\overline{\mathcal{G}}$ can be derived from $\mathcal{G}$, whose adjacency and Laplacian matrices are determined by

$$\overline{\mathcal{A}} = \begin{bmatrix} 0 & 1 & 0 \\ 1 & 0 & 0 \\ 0 & 0 & 0 \end{bmatrix} \text{ and } \overline{\mathcal{L}} = \begin{bmatrix} 2 & -1 & 0 \\ -1 & 1 & 0 \\ 0 & 0 & 1 \end{bmatrix}$$

Apparently, the subgraph $\overline{\mathcal{G}}$ is itself a directed graph.

Consider the \$i\$th follower agent $(i = 1, 2, 3)$. Take its control design of the x-axis subsystem into account. Some parameters of the integral SMC-based controller should be predefined, that is, $c_{xi} = 5$ and $\kappa_{xi} = 1$. Concerning the NDO-based observer, its gain vector is chosen as $\mathbb{L} = [0\ 6]^T$ by trial and error so that $\lambda_{xi} = \mathbb{L}^T\mathbb{B} = 6$. The constant ρ_{xi} in (7.17) is set by 1.1. The uncertain term of the x-axis subsystem is designed by $\delta_{xi} = 0.02 \times \text{rand()}$. Here rand() is a MATLAB command that generates a uniformly distributed random number in the closed interval $[-11]$. Without loss of generality, the integral SMC-based controller and the NDO-based observer of the y-axis subsystem are kept unchanged from those corresponding parameters of the x-axis subsystem. Consider the motor load of these follower agents, both u_{xi} and u_{ui} are confined to $u_{xi} \leq 0.5$ and $u_{yi} \leq 0.5$.

Consider the following formation task. The leader agent moves along a straight line. The follower agents keep tracking the leader and form up into a diamond-shaped formation. The straight trajectory of the leader is presented as follows. In a Cartesian coordinate system, the initial coordinates of the leader are located at [0m 0.6m]. Correspondingly, its velocities in the x-direction and y-direction are set by 0.2m⋅s^{-1} and 0.1m⋅s^{-1}, respectively. In order to form up into the desired diamond in this coordinate system, the initial coordinates of follower 1, follower 2 and follower 3 in order are placed at [0m 1.1m], [0m 0.8m] and [0m 0.3m], respectively. Their relative coordinations in order are assigned as [−0.2m 0.2m], [−0.4m 0m] and [−0.2m −0.2m] with respective to the leader agent.

Figure 7.4 displays the simulation results of the presented control scheme by the multi-agent system. In Figure 7.4a, the agents form up into the diamond-shaped formation from a string one while moving in straight lines, where those filled triangles denote the initial positions of the agents and those filled circles indicate the agents' positions in the dynamic process. In order to demonstrate the formation maneuver, the dash lines bond the agents together at the same moment. At the outset, the leader agent moves toward the right-half plane according to the designed trajectory. The three follower agents move towards the left-half plane in order to form up the desired diamond.

In Figure 7.4b-e, the position errors and the velocity errors of each follower agent in the x and y directions are illustrated. According to the designed communication topology in Figure 7.3, the position errors of follower 1 are defined by

$$\begin{aligned} e_{px1} &= [x_{h1} - (x_{h2} - d_{12}^{x})] + [x_{h1} - (x_{h4} - d_{14}^{x})] \\ e_{py1} &= [y_{h1} - (y_{h2} - d_{12}^{y})] + [y_{h1} - (y_{h4} - d_{14}^{y})] \end{aligned} \tag{7.33}$$

Similarly, we have

$$\begin{aligned} e_{px2} &= x_{h2} - (x_{h1} - d_{21}^{x}) \\ e_{py2} &= y_{h2} - (y_{h1} - d_{21}^{y}) \end{aligned} \tag{7.34}$$

$$\begin{aligned} e_{px3} &= x_{h3} - (x_{h2} - d_{32}^{x}) \\ e_{py3} &= y_{h3} - (y_{h2} - d_{32}^{y}) \end{aligned} \tag{7.35}$$

$$
\begin{aligned}
e_{vx1} &= v_{x1} - v_{x2} + v_{x1} - v_{x4} \\
e_{vy1} &= v_{y1} - v_{y2} + v_{y1} - v_{y4}
\end{aligned}
\tag{7.36}
$$

$$
\begin{aligned}
e_{vx2} &= v_{x2} - v_{x1} \\
e_{vy2} &= v_{y2} - v_{y1}
\end{aligned}
\tag{7.37}
$$

$$
\begin{aligned}
e_{vx3} &= v_{x3} - v_{x2} \\
e_{vy3} &= v_{y3} - v_{y2}
\end{aligned}
\tag{7.38}
$$

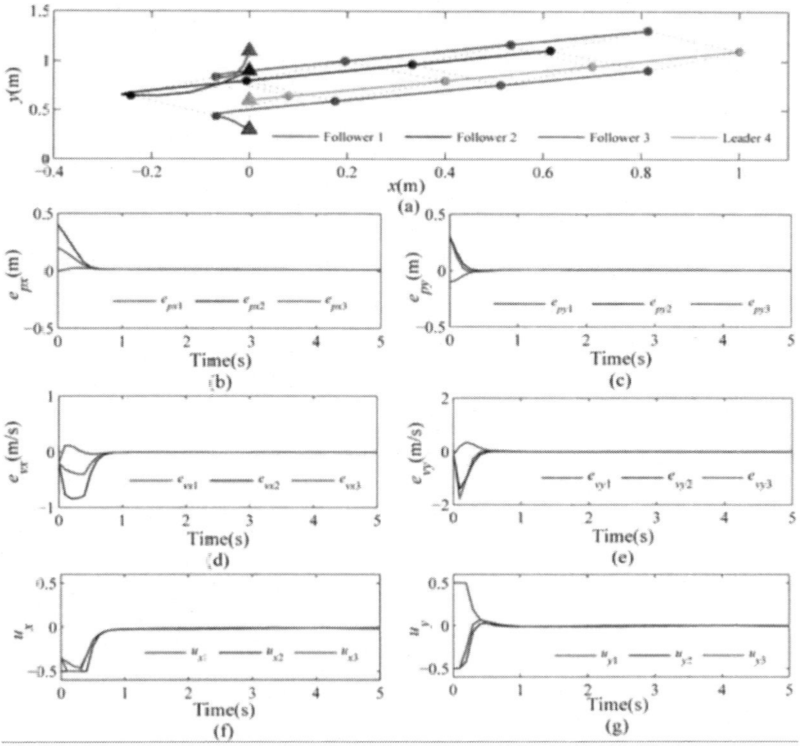

Figure 7.4. Simulation results of the control scheme. (a) Formation maneuvers in the Cartesian coordinate system, (b) Curves of e_{pxi}, (c) Curves of e_{pyi}, (d) Curves of e_{vxi}, (e) Curves of e_{vyi}, (f) Curves of u_{xi}, (g) Curves of u_{yi} ($i = 1, 2, 3$).

From Figure 7.4b-e, these defined errors can converge to zero as the desired formation has been achieved. The fact indicates the presented control scheme can realize the formation maneuver of the multi-agent system in spite of uncertainties. Further, the formation-control law of each follower agent is shown in Figure 7.4f-g.

CONCLUSION

This paper has investigated the formation-control problem of multiple agents. The agents under consideration are wheeled mobile robots. The formation mechanism is leader-following. The uncertainties originated from each individual agent result in the formation uncertainties of the multi-agent system. It is mildly assumed that the formation uncertainties are bounded by an unknown boundary. In order to resist the formation uncertainties when forming up the agents, a control scheme that integrates the technique of NDO-based observer and the method of integral SMC is addressed. According to a given communication topology, the theoretical analysis proves that the formation control of the multi-gent system in the presence of uncertainties is of asymptotic stability. The control scheme has achieved the formation maneuvers by a multi-robot platform. The simulation results have demonstrated the effectiveness of the control scheme.

This paper has presented some theoretical results, as well as some numerical results by the multi-robot platform. In order to focus on the motivation of control design, some difficulties in reality, such as communication delays and collisions between agents, are not considered during the control design. The no-communication-delay and no-collision conditions are mild enough for small-scale formations but they are rather idealized for large-scale formations. In order to take the presented control scheme into practical account, this field is our continuous interest and some contributions are still in progress.

REFERENCES

[1] Lin, P., Li, Z., Jia, Y., Sun, M. (2011). High-order multi-agent consensus with dynamically changing topologies and time-delays. *IET control theory & applications*, 5(8), 976-981.

[2] Su, Y., Huang, J. (2012). Cooperative output regulation with application to multi-agent consensus under switching network. *IEEE Transactions on Systems, Man, and Cybernetics, Part B (Cybernetics)*, 42(3), 864-875.

[3] Xiao, F., Wang, L., Chen, J., Gao, Y. (2009). Finite-time formation control for multi-agent systems. *Automatica*, 45(11), 2605-2611.

[4] Liu, C. L., Tian, Y. P. (2009). Formation control of multi-agent systems with heterogeneous communication delays. *International Journal of Systems Science*, 40(6), 627-636.

[5] Wu, Z., Guan, Z., Wu, X., Li, T. (2007). Consensus based formation control and trajectory tracing of multi-agent robot systems. *Journal of Intelligent and Robotic Systems*, 48(3), 397-410.

[6] Ni, W., Cheng, D. (2010). Leader-following consensus of multi-agent systems under fixed and switching topologies. *Systems & Control Letters*, 59(3-4), 209-217.

[7] Zhang, Z., Hao, F., Zhang, L., Wang, L. (2014). Consensus of linear multi-agent systems via event-triggered control. *International Journal of Control*, 87(6), 1243-1251.

[8] Li, Z., Duan, Z., Chen, G. (2011). Dynamic consensus of linear multi-agent systems. *IET Control Theory & Applications*, 5(1), 19-28.

[9] Luo, X., Han, N., Guan, X. (2011). Leader-following consensus protocols for formation control of multi-agent network. *Journal of Systems Engineering and Electronics*, 22(6), 991-997.

[10] Lu, X., Austin, F., Chen, S. (2012). Formation control for second-order multi-agent systems with time-varying delays under directed topology. *Communications in Nonlinear Science and Numerical Simulation*, 17(3), 1382-1391.

[11] Ren, W., Beard, R. W., Atkins, E. M. (2005, June). A survey of consensus problems in multi-agent coordination. In *Proceedings of the 2005, American Control Conference*, 2005. (pp. 1859-1864). IEEE.

[12] Gao, Y., Wang, L., Xie, G., Wu, B. (2009). Consensus of multi-agent systems based on sampled-data control. *International Journal of Control*, 82(12), 2193-2205.

[13] Gao, Y., Wang, L. (2011). Sampled-data based consensus of continuous-time multi-agent systems with time-varying topology. *IEEE Transactions on Automatic Control*, 56(5), 1226-1231.

[14] Dong, R., Geng, Z. (2015). Consensus for formation control of multi-agent systems. *International Journal of Robust and Nonlinear Control,* 25(14), 2481-2501.

[15] Zuo, Z., Tie, L. (2014). A new class of finite-time nonlinear consensus protocols for multi-agent systems. *International Journal of Control*, 87(2), 363-370.

[16] Yu, W., Chen, G., Cao, M., Ren, W. (2013). Delay-induced consensus and quasi-consensus in multi-agent dynamical systems. *IEEE Transactions on Circuits and Systems*, 60(10), 2679-2687.

[17] Defoort, M., Polyakov, A., Demesure, G., Djemai, M., Veluvolu, K. (2015). Leader-follower fixed-time consensus for multi-agent systems with unknown non-linear inherent dynamics. *IET Control Theory & Applications*, 9(14), 2165-2170.

[18] Hu, J., Feng, G. (2010). Distributed tracking control of leader–follower multi-agent systems under noisy measurement. *Automatica*, 46(8), 1382-1387.

[19] Du, H., Wen, G., Wu, D., Cheng, Y., Lü, J. (2020). Distributed fixed-time consensus for nonlinear heterogeneous multi-agent systems. *Automatica*, 113, 108797.

[20] Anggraeni, P., Defoort, M., Djemai, M., Zuo, Z. (2019). Control strategy for fixed-time leader–follower consensus for multi-agent systems with chained-form dynamics. *Nonlinear Dynamics*, 96(4), 2693-2705.

[21] Ning, B., Han, Q. L., Zuo, Z. (2019). Practical fixed-time consensus for integrator-type multi-agent systems: A time base generator approach. *Automatica*, 105, 406-414.

[22] Oh, K. K., Park, M. C., Ahn, H. S. (2015). A survey of multi-agent formation control. *Automatica*, 53, 424-440.

[23] Young, K. D., Utkin, V. I., Ozguner, U. (1999). A control engineer's guide to sliding mode control. *IEEE transactions on control systems technology*, 7(3), 328-342.

[24] Levant, A. (1993). Sliding order and sliding accuracy in sliding mode control. *International Journal of Control*, 58(6), 1247-1263.

[25] Li, S., Yu, X., Fridman, L., Man, Z., Wang, X. (Eds.). (2017). *Advances in Variable Structure Systems and Sliding Mode Control—Theory and Applications* (Vol. 115). Springer.

[26] Bai, J., Wen, G., Rahmani, A., Yu, Y. (2017). Distributed consensus tracking for the fractional-order multi-agent systems based on the sliding mode control method. *Neurocomputing*, 235, 210-216.

[27] Han, S. I. (2017). Prescribed consensus and formation error constrained finite-time sliding mode control for multi-agent mobile robot systems. *IET Control Theory & Applications*, 12(2), 282-290.

[28] Wang, G., Wang, X., Li, S. (2018). Sliding-mode consensus algorithms for disturbed second-order multi-agent systems. *Journal of the Franklin Institute*, 355(15), 7443-7465.

[29] Ren, C. E., Chen, L., Chen, C. P., Du, T. (2016). Quantized consensus control for second-order multi-agent systems with nonlinear dynamics. *Neurocomputing*, 175, 529-537.

[30] Jiang, Y., Liu, J., Wang, S. (2016). Robust Integral Sliding-Mode Consensus Tracking for Multi-Agent Systems with Time-Varying Delay. *Asian Journal of Control*, 18(1), 224-235.

[31] Chen, S., Wang, M., Li, Q. (2020). Second-Order Consensus of Hybrid Multi-Agent Systems With Unknown Disturbances Via Sliding Mode Control. *IEEE Access*, 8, 34973-34980.

[32] Mirabdollahi, S. E., Haeri, M. (2019). Multi-agent system finite-time consensus control in the presence of disturbance and input saturation

by using of adaptive terminal sliding mode method. *Cogent Engineering*, 6(1), 1698689.

[33] Mishra, R. K., Sinha, A. (2019). Event-triggered sliding mode based consensus tracking in second order heterogeneous nonlinear multi-agent systems. *European Journal of Control*, 45, 30-44.

[34] Zhou, J., Zeng, D., Lu, X. (2020). Multi-Agent Trajectory-Tracking Flexible Formation via Generalized Flocking and Leader-Average Sliding Mode Control. *IEEE Access*, 8, 36089-36099.

[35] Chen, W. H., Yang, J., Guo, L., Li, S. (2015). Disturbance-observer-based control and related methods—An overview. *IEEE Transactions on Industrial Electronics*, 63(2), 1083-1095.

[36] Kempf, C. J., Kobayashi, S. (1999). Disturbance observer and feedforward design for a high-speed direct-drive positioning table. *IEEE Transactions on control systems Technology*, 7(5), 513-526.

In: A Closer Look at Formation Control
Editor: Dianwei Qian
ISBN: 978-1-53618-177-7

Chapter 8

A Fuzzy Logic-Based Adaptive Strictly Negative-Imaginary Formation Controller for Multi-Quadrotor Systems

Vu Phi Tran*, Fendy Santoso and Matthew A. Garratt
School of Engineering and Information Technology,
University of New South Wales, Canberra, Australia

Abstract

Multi-robot systems have played essential roles in both civilian and military domains. The task assigned to the robotic systems has become more complicated and time-consuming to be carried out by a single robot. To reduce the costs and to improve the robustness and the efficacy of the system, formation control approaches for networked robots are highly desirable e.g., for a team of Unmanned Aerial Vehicles (UAVs). However, current state-of-the-art algorithms are faced with several challenging issues due to the presence of uncertainties such as wind gusts. Since traditional PID controllers or other fixed-gain control systems may

* Corresponding Author's Email: phivuspkt@gmail.com.

not lead to satisfactory outcomes, we present a novel adaptive Strictly Negative Imaginary (SNI) formation control strategy based on knowledge-based Fuzzy Inference Systems to deal with the challenging dynamics of flight environments. To facilitate the information sharing process, a decentralised control architecture is designed. Each UAV only needs to measure its relative position with respect to its neighbours with a sensor such as a laser range-finder or machine vision system. Consequently, the required formation is formed by maintaining the desired relative positions between UAVs. We rigorously compare the performance of our adaptive SNI control systems with respect to the effectiveness of the conventional PID controllers. Our research highlights the efficacy of the proposed adaptive control systems, not only using numerical simulations, but also by conducting real-time flight tests.

Keywords: distributed formation control, adaptive strictly negative-imaginary controller, networked robots, fuzzy systems, unmanned aerial vehicles, adaptive control

INTRODUCTION

Formation control of networked multi-robot systems (MRS) has drawn significant attention in recent years due to state-of-the-art applications in many areas, including both civil and military domains. The general formation control problem can be defined as the inter-relational physical structure in which a group of autonomous vehicles cooperates in the network to form and maintain a tightly pre-defined spatial pattern while tracking a given trajectory. Applications of formation control in the literature include firefighting [1], securing a protected region, surveillance [2, 3], navigation and planning [4].

Multiple robots moving in a formation possess many benefits, compared to single robot systems. For instance, they can bring system costs down and upgrade the robustness and effectiveness of a system while creating higher redundancy, reconfiguration capability and structural flexibility. Satellites orbiting in formation can expand their target sensing capabilities [5,6]. In an automated highway system, self-driving vehicle formations can prevent congestion as well as improve road network efficiency [7]. In mapping and planning applications, a formation method allows the autonomous movements of a large-scale MRS to be managed and controlled efficiently. In summary, the formation control for networked MRS benefits from improving system performance and reducing completion time for complex tasks.

The selection of MRS control protocols depends on the form of coordinated manner and each robot's capabilities to accomplish a given task. Most of today's control architectures fall into two categories: centralised and decentralised. In centralised structures, a central station, which could be either a computer or a robot, acquires all required data including the global information of an environment and the states of the whole team before publishing proper control signals to each robot, until they arrive at the goal point [8, 9]. However, the original issue in centralised computation typically arises once a large team of robots is involved in cooperative control. Since there is only one centralised processor to control all of the systems, the server can fail due to computational overload or information sharing within the entire team can be occasionally disrupted due to limited communication or sensing power resources. These serious problems can lead to unsatisfactory performance or even task failure. Since advancing technology has rendered embedded processors for autonomous vehicles more powerful and inexpensive, the distributed control system has gradually replaced the centralised model in most of the applications of intelligent autonomous robots [10-13]. Each robot, accompanied by a local controller generates its own control by updating its local states and its neighbors' states. Consequently, distributed processing significantly reduces the computational power and the amount of information sharing among robots, directly contributing to the reliability and robustness against communication failures.

Having a similar concept, under the leader-follower (L-F) strategy, one robot in the formation, assigned the role of the leader, defines the path for entire formation while the follower robots position themselves with respect to the position and orientation of the leader during movement. This leads to cost reduction and simplicity in the formation generation since all robots do not synchronise their own states, and the follower robots do not require global information. Consequently, a large variety of L-F formation control approaches have been investigated extensively in the past decade [14, 15]. However, if the leader robot is lost mainly due to additional constraints (e.g., sensor limitations or the occlusion of line-of-sight visibility by obstacles) or encounters a system failure (e.g., communications breakdown or hardware problems), the formation stabilisation and the entire system, made by multiple robots, can be no longer maintained.

On the other hand, under the virtual-leader strategy, all member vehicles of the formation are considered as the followers. Their task is to maintain a relative position from a simulated leader regarding a formation trajectory. The virtual structure method is a typical virtual-leader strategy. To implement this control technique, formation trajectory planning, global position measurement and information synchronization must be implemented on all robots. These requirements aim at providing system robustness against any fault or disappearance of team members. Nevertheless, the virtual-leader strategy consumes a high cost and considerably computational complexity since all robots simultaneously acquire the same information.

To obtain robustness against the loss of the leader, distributed formation tracking control solutions based on relative position consensus have become a developing trend in multi-mobile robot formation control. However, many proposed formation control algorithms [16-18] have mainly focused on the consensus of first-order or second-order multi-robot systems, whereas the UAV quadrotor or UGV systems should be represented by higher-order flight dynamics models [19-21]. Despite the need to overcome these problems, only a few mature and sophisticated approaches for solving the consensus problem in ideal environments have been proposed in the literature. These techniques include adaptive methods [22], sliding mode control [23, 24], and potential field function [25, 26]. In outdoor applications, uncertain factors like wind gusts can severely deteriorate the performance of aerial robots when they are flying at close range to one another.

Extensive works have been done to solve this problem. The existing disturbance rejection methods can be classified into two strategies:

(a) wind gusts model [27, 28] and (b) adaptive controllers [29]. The former measures the unknown wind forces and directions, develops a wind gust model via system identification, and then cancels it in the controller structure. However, the wind gust models are often inaccurate since it is challenging to measure the wind direction. The latter, which promises to show a better performance, has not yet been extensively studied. The controller parameters have been adjusted in such a way to adapt to environmental changes, for example, increasing the damping and DC gain factors. By doing this way, the unexpected costs for wind sensor and modeling can be significantly reduced.

Compared to other methods considering the distributed formation control problem, we leverage the advantages of a Strictly Negative Imaginary (SNI) consensus controller that can provide robust stability results. As discussed in [30, 31], the theory guarantees an internal stability for the feedback interconnection of the two systems: a Negative Imaginary (NI) plant and the SNI controller whose DC gain at the zero frequency is less than unity. Based on this feature, a distributed consensus algorithm using global information was developed for some ideal SNI systems and tested in simulation [32, 33]. After that, a NI consensus control law-based the relative position information was implemented on two real UAV systems in order to synchronize their behavior with a offset distance [13]. Nevertheless, in this earlier work, the NI consensus controller was unable to cope with environmental uncertainties. Hence, to the best of our knowledge, our research group is among the first in the literature that presents a practical approach to the distributed multi-UAV formation control problem without a leader flying in disturbed environments. With this technique, our aerial vehicles can regulate the relative positions to achieve and maintain a desired formation through a communication topology when compensating for the unexpected effects of wind gust disturbances. In both cases, we also perform a comparative study regarding the performance of conventional PID controllers.

Furthermore, many autopilot systems applied in unmanned aerial systems still heavily depend on the benefits of the fixed-gain controller due to their simplicity [13, 34], whereas, in outdoor activities, most process dynamics are intrinsically time-varying and the UAV flights are always subjected to the uncertainties such as unpredictable wind gusts, ground effects, atmospheric conditions and battery charge [35]. Addressing this circumstance, a fuzzy-SNI controller, which is constructed by the techniques of the fuzzy logic system and the SNI controller, is proposed in this work to achieve a high-performance control system in the presence of multiple disturbances. The knowledge based-fuzzy inference system is employed to facilitate automatic tuning in our SNI controller, improving its level of robustness. As the language of uncertainty, fuzzy systems are integrated with many other techniques (e.g., PI-type control [36], disturbance-observer-based control [37] and a composite controller [38]) to solve the anti-disturbance tracking-control problems for a class of single input-single output (SISO) or multi input-multi output (MIMO) systems with uncertain disturbances [39-41]. To date, there is no single research paper discussing the effectiveness and robustness of the fuzzy logic to self-tune the SNI autopilot.

In this chapter, we proceed as follows. While the introduction part is described in this section, Section II provides some essential background on physical platforms and communication systems used to conduct various cooperative tasks. Meanwhile, Section III delineates the UAV mathematical modeling. A detailed study of the existing NI-systems algorithms, which includes consensus control and formation behaviour, is introduced in Section IV to provide the theoretical foundation for further innovation. Our novel formation control theory based on the adaptive SNI consensus controller and the Takagi-Sugeno Fuzzy Inference System (FIS) tuning is investigated in Section V. Section VI and VII are used to highlight the efficacy of our distributed adaptive formation control method with respect to the performance of conventional PID controller, widely used for many aerial robotics applications. Lastly, Section VIII concludes this chapter.

UNMANNED AERIAL VEHICLES: HARDWARE AND SOFTWARE

UAVs with diverse capabilities are starting to play essential roles in many aspects of modern life. The broad spectrum of applications of the UAVs led to the invention of various types of UAV platforms, namely, fixed-wing, rotary-wing and wing aircraft. While fixed-wing aircraft are more suitable for the long flight missions at high speed, rotary wing aircraft, especially quadcopter drones, have a smaller range and lower cost than fixed-wing aircraft. They are often employed in indoor applications or in the cluttered environments or in tiny spaces due to their lower flight speed, vertical take-off and landing capability, in addition to their ability to hover.

Figure 8.1. A hovering AR.Drone 2.0 quadrotor. The system attitudes is expressed as $[\phi\ \theta\ \psi]^T$, which represents the angles of roll, pitch, and yaw, respectively.

We select the AR.Drone 2.0 quadrotor for our experiments due to its lightweight construction (450 grams), resulting in greater sensitivity to gusts of wind than other platforms. Thus, this type of drone requires a high-performance tracking controller design. Our AR.Drone 2.0 quadrotor (see Figure 8.1) is equipped with two cameras (forward and downward looking). Our AR.Drone 2.0 quadrotor has two on-board 720p high definition cameras (forward and downward looking) located in a underneath indoor hull with rotor guards, besides an ultrasound sensor to measure the UAV altitude with respect to the ground. The drone is naturally crash tolerant, making it appropriate to both indoor and outdoor tasks. The aircraft employs a Wi-Fi network with a line-of-sight range of up to 50 meters so that it can be remotely controlled through a smart device or central computer.

Due to the lack of on-board micro-processor and peripheral interface modules, the AR.Drone control programs can be written in a central computer using the AR.Drone autonomy package developed in the Robot Operating System (ROS) environment. The control outputs, including the rate of pitch, roll, yaw, and thrust commands, are then sent from the central station to the relevant AR.Drone quadrotor using the drone's IP address at a frequency of 100 Hz. In our multi-AR.Drone scheme, each vehicle is supervised and controlled by a computer, whose function is similar to the on-board micro-processor. This setting enables formation control applications to be implemented in a distributed manner as shown in Figure 8.2.

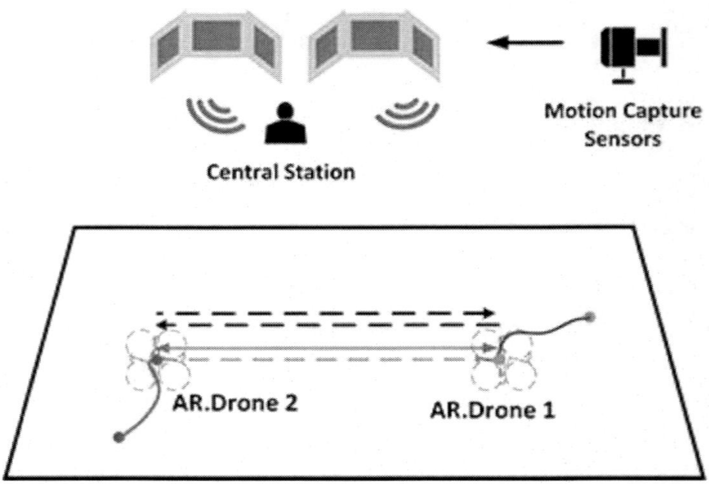

Figure 8.2. Distributed formation control of two AR.Drone quadcopters.

SYSTEM IDENTIFICATION

The availability of accurate and effective dynamics models plays an important role in the performance of model-based control systems. In Section II, system identification is conducted based on the fundamental principles of the black box, where no prior model of the plant is available. This means that the whole closed-loop system's behavior, considered as a plant, can be represented in terms of transfer functions (the mathematical models between input and output signals).

It is assumed that the quadrotor dynamics are inherently non-linear, whereas, the SNI systems require a linear model. Hence, we will linearise the non-linear dynamics of our quadrotor drone along its x, y, and z axes, which is valid within the normal operating points of the system. To do so, the drone's roll and pitch motion is constrained at $\pm 30^{\circ}$ and the maximum velocities on the planar plane are 0.2 m/s.

Further, to collect the necessary data for modeling, some initial work must be carried out. Firstly, the attitude motion and translational velocity of the quadrotor on the three directions were stabilized by the well-tuned PID controllers. Next, the desired velocity inputs on the three axes were changed from lower to higher frequencies to capture the overall dynamics of the drone during flight tests. Meantime, all real flight data was acquired and stored in a ground station (GS).

As soon as the data collected was sufficient for the model estimation, we began to build up the system model and evaluation metrics. Since the experiment-based system identification technique has been proposed in the literature to find out the AR.Drone quadcopter's transfer functions [12], the Auto-regressive Moving Average Exogenous (ARMAX) model was selected due to its popularity. Moreover, its accurate level was validated through fit percentage tool between the actual output and the predicted output.

Given input and output signals, a single-input, single-output system identification using the ARMAX model and the fit percentage equation can be formulated as follows:

$$A\left(z^{-1}\right)y = z^{-k}B\left(z^{-1}\right)u + e \tag{8.1}$$

$$fit\left(\%\right) = \left(1 - \frac{y - \tilde{y}}{y - \text{mean}\left(y\right)}\right)*100\% \tag{8.2}$$

where z denotes the one step forward operator, while u is the system input and y is the system output. Further, k highlights the time step. $A(z^{-1})$ and $B(z^{-1})$ indicate the matrix real polynomials and e is the modeling error. y and $\tilde{y}$ are the validation data output and the output of our real UAV system respectively.

The velocity loop transfer functions of our AR.Drone quadcopter along the x, y and z axes are given by

$$P_x\left(s\right) = \frac{x\left(s\right)}{\dot{x}\left(s\right)} = \frac{-0.62s^3 - 866.3s^2 + 3156.3s + 39427.5}{s^4 + 31s^3 + 53425.2s^2 + 17296.3s + 1093.7} \tag{8.3}$$

$$P_y\left(s\right) = \frac{y\left(s\right)}{\dot{y}\left(s\right)} = \frac{-1.71s^3 + 249.1s^2 - 66651.5s + 68465.9}{s^4 + 78.9s^3 + 58525.8s^2 + 50869.1s + 970.1} \qquad 8.4)$$

$$P_z\left(s\right) = \frac{z\left(s\right)}{\dot{\tilde{z}}\left(s\right)} = \frac{-0.026s^2 + 0.3}{s^2 + 2.25s + 0.019} \tag{8.5}$$

where ($x(s)$, $y(s)$, $z(s)$) and ($\dot{x}\left(s\right), \dot{y}\left(s\right), \dot{\tilde{z}}\left(s\right)$) indicate the position outputs and their corresponding velocity set points along each axis respectively.

Based on the obtained transfer functions in (8.3)–(8.5), the accuracy was calculated using (8.2). From Figure 8.3, it is clear that both quadcopter velocity transfer functions have a reasonably small modeling error of about 1%, meaning that our models were able to predict AR.Drone quadcopter behaviors accurately within given constraints.

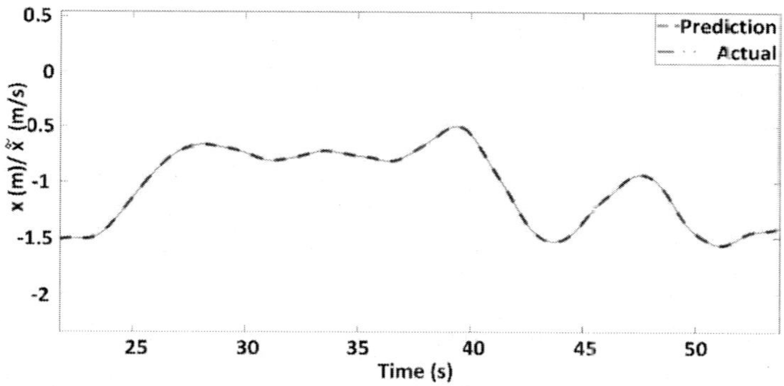

(a)

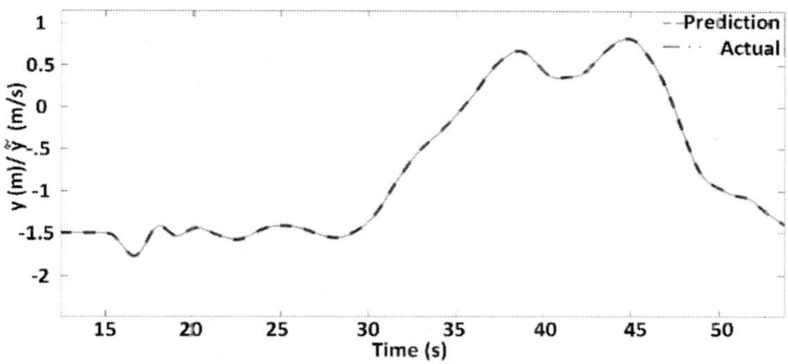

(b)

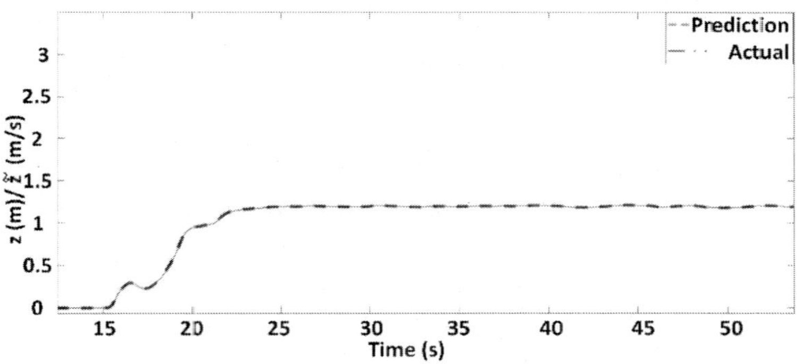

(c)

Figure 8.3. Prediction accuracy of system performance in modeling.
(a) X-Axis, (b) Y-Axis and (c) Z-Axis

NI Systems Theory

The SNI/NI system class has attracted growing attention among control theory researchers since its formal definition was proposed in [42], after which it speedily gained a variety of successful applications in aerospace, large space structures [43], multi-link robotic arms [44], and nano-positioning control wherein collocated position measurements and force actuators [44, 45] are typical. NI systems theory can also be employed to address other areas of advanced technique. Importantly, such unstable systems whose dynamic models are delineated by a string of subsystems (e.g., beam vibration, flutter instabilities and large platoons of vehicles) can be controlled by an NI feedback-control system, as in [46]. In this section, a survey of great achievements in both theory and applications of SNI/NI systems was introduced.

SNI/NI Systems Theory

For a general system whose transfer function $P(s)$ has NI frequency response, the SNI property is defined by the following requirements:

1) $P(s)$ possesses no poles at the origin and any pole's real part at the right-half plane.
2) For the positive frequencies, its Nyquist contour is always located below the real axis. The mathematical condition for a SNI system can be described as follows: $N(s) \in \text{SNI}$ if $j[N(j\omega) - N^*(j\omega)] > 0$ for all $\omega > 0$.
3) Likewise, based on a graphical definition, if the system's Nyquist plot begins from an arbitrary point on the real axis, while the frequency ω is 0 and ends at 0, whereas the frequency ω is ∞ (see Figure 8.4), it surely contains SNI property. Moreover, the Nyquist figure never encircles the critical point (1, $j0$), meaning that the SNI system has robust stability.

4) The transfer function matrix $N(s) = C(sI - A)^{-1}B + D$ is such that $N(s) - N^T(-s)$ has no transmission zeros on the imaginary axis, except possibly at s=0. Here, A, B, C and D represent the state, input, output and feed-forward matrices, respectively.

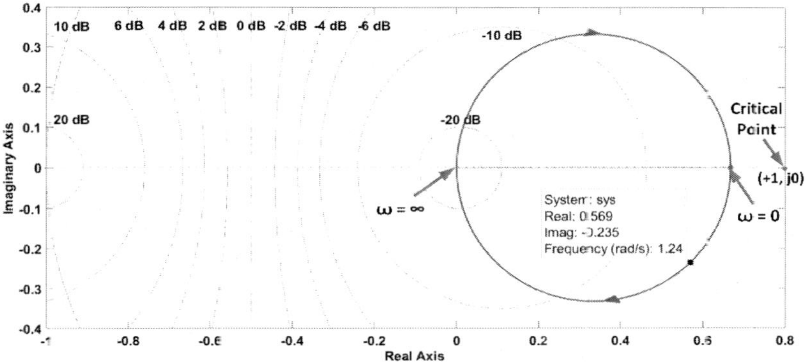

Figure 8.4. Nyquist plot of a SNI transfer function $P(s) = \frac{2}{s+3}$, which indicates that the imaginary term of $P(j\omega)$ is negative for all $\omega > 0$.

Similarly, a transfer function is treated as a NI system if the following conditions are met:

1) All poles locates in the open left half s-plane.
2) Its Nyquist figure lies below the real axis for the positive frequencies and touch, but does not cross the real axis as shown in Figure 8.5. This condition can be mathematically expressed as follows: $N(s) \in \text{NI}$ if $j[N(j\omega) - N(-j\omega)] \geq 0$ for all ω>0.

Based on the definition of a SNI/NI system, we constructed the experimental Nyquist plots of [$P_x(s)$ $P_y(s)$ $P_z(s)$]. Following Figure 8.6, the Nyquist plots' shape satisfies the third condition of SNI systems. Hence, all AR.Drone quadcopter transfer functions are SNI.

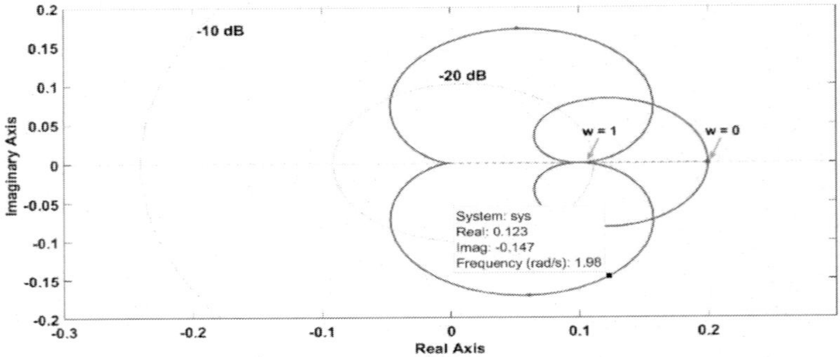

Figure 8.5. Nyquist plot of a NI plant $P(s) = \frac{2s^2+s+1}{2s^4+7s^3+17s^2+17s+5}$, where the imaginary part of $P(j\omega)$ is negative for all $\omega \geq 0$, except $\omega=0$ and $\omega=1$. Since the imaginary part of $P(j\omega)$ is zero, $P(s)$ must be NI, but not SNI.

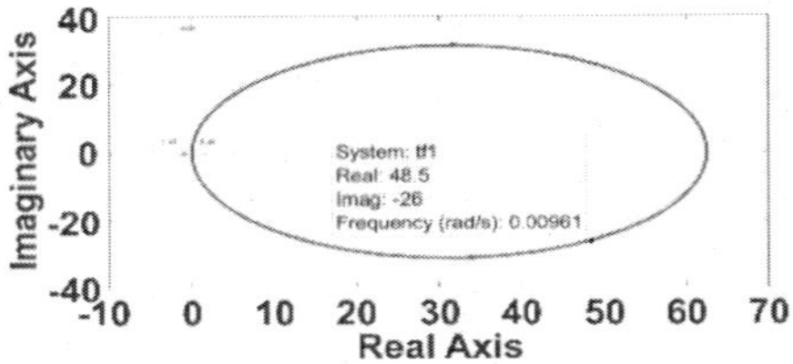

(a)

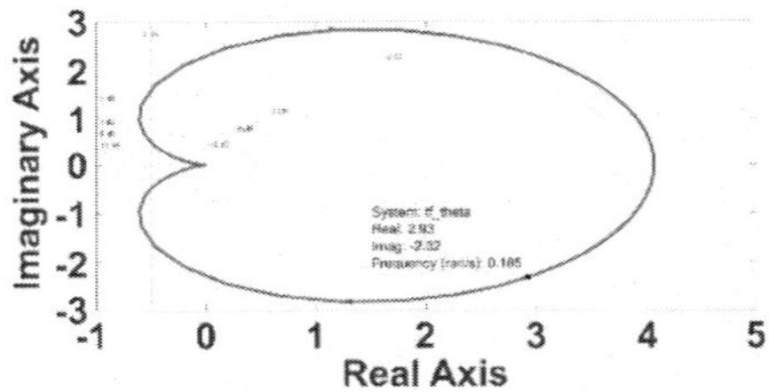

(b)

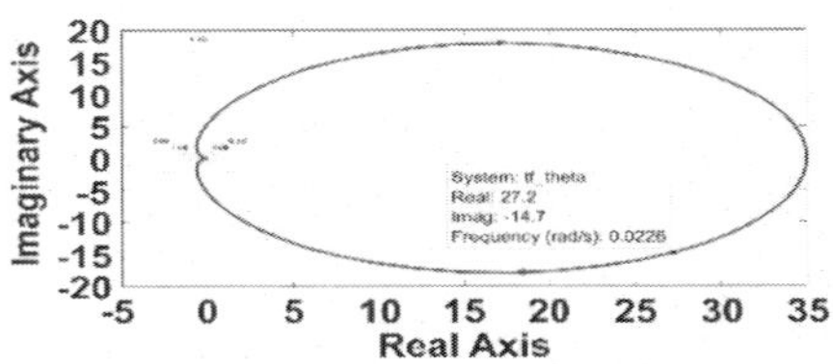

(c)

Figure 8.6. Nyquist plots of the vertical and horizontal AR.Drone quadcopter plants. (a) X-Axis, (b) Y-Axis and (c) Z-Axis.

Graph Theory

Typically, the accomplishment of cooperative tasks relies on effective information interchange and collaboration between team members. Recent works in terms of consensus control and state agreement have been closely related to multi-hop communication for convergence and performance guarantees. Networks possess long served as models of local intercommunication in the field of mobile robotics where robots are typically associated with the nodes of a graph and communication links including the edges.

Graphs are constructed to graphically represent pairwise relations between a finite number of robots by means of vertices and edges. Alternatively, graphs also admit representations in terms of numerical matrices. Essentially, a graph $G=(V, E)$ is composed of a set of n nodes $V=\{v_1, v_2, \ldots, v_n\}$ which are linked by j edges $E\subseteq(V\times V)$. For directed graphs, edges can be regarded as a direction from one node to another $e=(u, v)\in E$. Undirected graphs are a special case of directed graphs where (u, v), (v, u): $<u, v> \in E$. The incidence matrix for G with respect to an ordered set $\{v_1, v_2, \ldots, v_n\}$ is an $n\times l$ matrix Q, given by:

$$Q=\begin{cases} q_{ij}=1 & \text{if } i \text{ is the inital vertex of edge } j \\ q_{ij}=-1 & \text{if } i \text{ is the terminal vertex of edge } j \\ q_{ij}=0 & \text{if } i \text{ is not connected to } j, (v_i, v_j)\notin E \end{cases} \tag{8.6}$$

in which the Laplacian matrix $\mathcal{L}_n$ is defined as $\mathcal{L}_n=QQ^{\mathrm{T}}$. Further, it is interesting to see that $\mathcal{L}_n$ has the following properties when the graph is undirected and connected:

$$\mathcal{L}_n \geq 0, \mathrm{Ker}(\mathcal{L}_n)=\mathrm{span}\{1_n\}. \tag{8.7}$$

Fixed-Gain NI Formation Control Strategy Based on the Relative Position Consensus

Although the NI-systems control area has been particularly active over the past decade, a single SNI system cannot address complicated and time-consuming tasks such as exploration and mapping, heavy load transportation system, or delivery services. These tasks require coordination and information sharing among multiple SNI systems. Thus, more researchers are focusing on the robust cooperative control strategies for multi-SNI systems. For example, a distributed output-feedback consensus method based on the global information and a fixed-communication topology was designed for heterogeneous multi-input, multi-output (MIMO) NI plants [32, 33]. In the theory, all networked NI systems share the output feedback information and naturally converge a rendezvous point, even when both input and output channels were encountering multiple perturbations.

We now impose the following assumptions throughout the rest of this paper:

Assumption 1: Graph $\mathcal{G}$ is always undirected and connected.

Assumption 2: $diag_{i=1}^{n} P_i(s)$ is SNI if and only if each $P_i(s)$ is SNI $\forall i \in \{1, 2, \ldots, n\}$ agents.

Assumption 2 can be straightforwardly proven from the definition of the SNI systems since each SNI system's model is located diagonally in an SNI-systems matrix. The same argument is also applied to NI functions.

Given multiple SNI systems with n>1 agents, arranged in a diagonal matrix $\overline{P}(s)$ of n×n, and given SNI-consensus-controllers matrix $\overline{A}(s)$, arranged in the same dimension as $\overline{P}(s)$, the output feedback consensus for each agent to reach a rendezvous point under the information communication graph, namely, Laplacian matrix $\mathcal{L}_n$, is expressed as follows (see Figure 8.7).

$$\bar{y} = \bar{P}(s)\bar{u} \tag{8.8}$$

$$P_r = 1_n \otimes r \quad (8.9)$$

$$\bar{u} = \bar{A}\mathcal{L}_n(P_r - \bar{y}) \tag{8.10}$$

where ⊗ represents the Kronecker product, while $\bar{u}$ and $\bar{y} \in \mathbb{R}^{n\times n}$ are the input and output matrices of $\bar{P}(s)$. $r\in \mathbb{R}^{3\times 1}$ turns out the master UAV's reference position along the three directions. P_r denotes the desired consensus position or convergence point matrices of the networked agent team. Moreover, in the general consensus control scenario of the networked systems, (8.10) is equivalent to the following protocol:

$$u_i = A_i(s)\sum_{j=1}^{n} q_{ij}(y_i - y_j) \tag{8.11}$$

where a_{ij} is the elements of the incidence matrix, $u_i\in \mathbb{R}^{n\times 1}$ and $y_i\in \mathbb{R}^{n\times 1}$ are the input setpoint and actual output of each agent member.

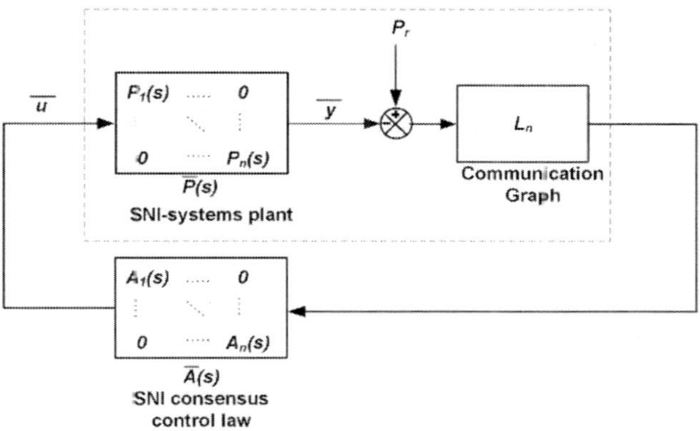

Figure 8.7. Distributed consensus control structure based on SNI-systems theory.

The output feedback consensus between agents is achieved if and only if the following condition is fulfilled

$$\lim_{t\to\infty}\sum_{i=1}^{n}\left(y_i - y_{ss}\right)=0 \tag{8.12}$$

where y_{ss} is the final rendezvous point.

A direct application of NI cooperative tracking, robust rendezvous for the three AR.Drone quadcopters was implemented in a 3D scenario in order to validate the convergence of the proposed method. We employed UAV plants that were developed in Section III. The pre-defined rendezvous point P_r is [60 −45 100]. All 3D SNI consensus controllers in $\overline{A}(s)$ matrix are selected as $[-\frac{s+300}{s+2} \quad -\frac{s+270}{s+2} \quad -\frac{s+11}{s+2}]$. The initial positions are [62 −50 0], [−68 50 0], [−8 −50 0] respectively. The graph is exactly the same as Figure 8.8 and the trajectory planner only sends information to the drone master (8.1) while the follower 2 (8.2) is treated as the master of the follower 3 (8.3), which produces the Laplacian matrix $\mathcal{L}_n = \begin{bmatrix} 2 & -1 & 0 \\ -1 & 2 & -1 \\ 0 & -1 & 1 \end{bmatrix}$.

Figure 8.8. Communication graph for three drones.

From Figure 8.9, it can be seen that robust rendezvous for the three UAV plants is obtained after only 12 s using NI cooperative tracking controller. Three AR.Drone drones, located at different positions, are coordinated under the directed graph topology and controlled to reach a flight consensus at the final rendezvous point when the position errors are 0. More importantly, there are three additional conclusions drawn from the obtained results. Firstly, since all aerial-vehicle trajectories simultaneously converge on a reference position, which is leading to a serious collision between robots, a formation control strategy, generating the relative positions between drones, is required to safely drive the UAV systems in a given formation. Secondly, the position output $y \in \mathbb{R}^{9\times1}$ obtains consensus $(\sum_{j=1}^{3\times n} y_i = y_j \quad i \neq j)$ when $\overline{y} \to 0 \in \mathbb{R}^{9\times1}$. As a result, the position output consensus problem can be directly converted to an internal position stability. Finally, the position output consensus among multiple AR.Drones is achieved using the global information. Hence, it is not appropriate for the distributed formation control.

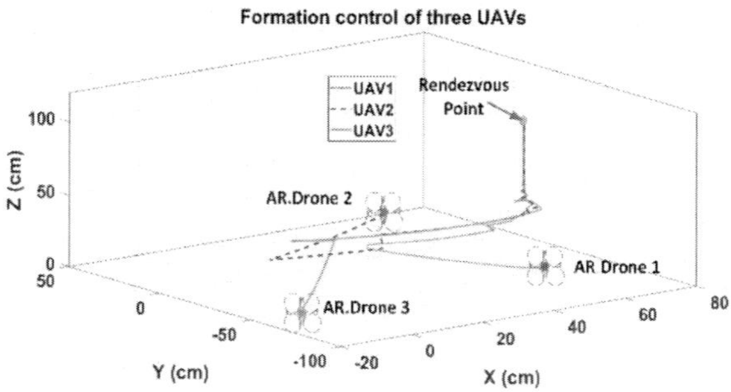

a

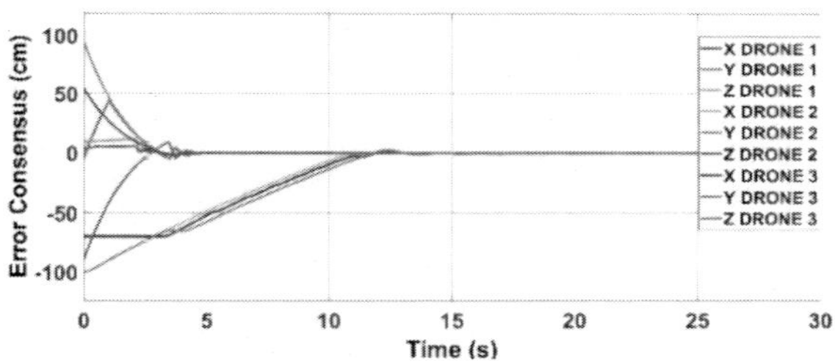

b

Figure 8.9. Robust rendezvous for the three UAV plants.
(a) 3D Consensus paths of the three AR.Drone quadcopters; (b) Position consensus errors along three axes.

We recall an initial research outcome within our research group in [13], where the distributed formation control based on a fixed-gain NI relative position consensus controller was designed and tested in simulation and real-time flight experiments. The relative state information between the UAVs was amended to the original structure in [32]. Based on Figure 8.10, the final form of the NI formation control strategy becomes as follows:

$$P_d = 1_n \otimes d \tag{8.13}$$

$$\bar{d} = \mathcal{L}_n \left(P_r - \bar{y} \right) \tag{8.14}$$

$$\bar{u} = \bar{A}\left(P_d - \bar{d} \right) \tag{8.15}$$

where $d \in \mathbb{R}^{3\times 1}$ indicates the desired relative positions between the master UAV and the followers UAV in a formation team.

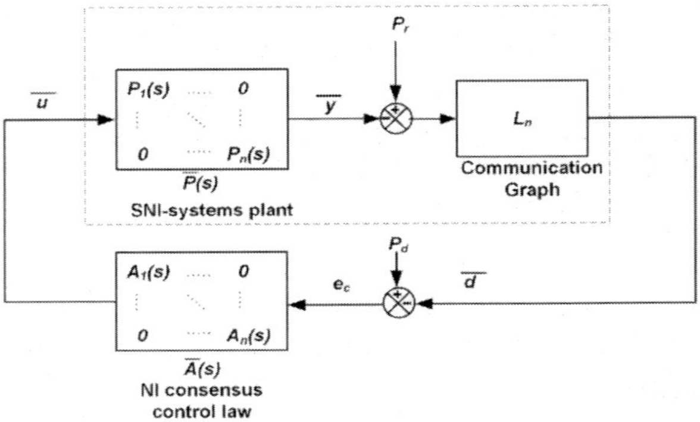

Figure 8.10. Distributed consensus control structure based on SNI-systems theory.

Although our research outcome clearly indicates a potential control method since the follower UAV was able to mimic the master UAV's behavior in tracking a rectangular trajectory, the real flight paths were basically oscillating around our baseline. This result comes from the turbulent airflow between the two UAVs during the real tests. Assuming that the cooperative performance of multi-UAV systems was disturbed by $\omega \in \xi$ perturbing factors (e.g., wind gusts or ground effect), it is clear that $\bar{y} \to P_r + \omega$, $P_d \to \bar{d} + \omega$ and the consensus errors $e_c \to \bar{d} + \omega$ when the consensus between agents is achieved. Moreover, the fixed-gain NI formation controller has not an anti-disturbance characteristic, for example, increasing the damping ratio for the whole system. To solve this problem, we will develop an adaptive SNI formation controller based on the fuzzy logic rules.

ADAPTIVE SNI FORMATION CONTROL APPROACH USING FUZZY LOGIC

Negative Imaginary (NI) systems theory has been applied to other areas of advanced technique. In [45-47], NI robust controllers were designed to stabilise high-precision systems such as atomic force microscopes, nano-positioners, micro-robotics and hard disks in the presence of multiple uncertainties (e.g., unmodelled dynamics, sensor noise and actuator faults). In this chapter, its application has been extended to external disturbance rejection for UAV formation flights.

SNI Relative-Position Consensus Controller

The definition of a single input-single output system that satisfies the SNI property was firstly presented in 2008 [42]. The necessary and sufficient conditions for a general transfer function N(s) to become SNI are given as follows:

1) $N(s)$ has no poles at the origin and all its poles have negative real parts.
2) The Nyquist plot of $N(s)$ is always located below the real axis for all positive frequencies as mathematically described as follows: $N(s) \in \text{SNI}$ if $j[N(j\omega) - N(-j\omega)] > 0$, $\forall \omega > 0$.
3) The Nyquist plot of an SNI system start from an arbitrary point on the real axis while the frequency ω is 0 and ends at 0 while the frequency ω is ∞. This means that the Nyquist of the system shall never encircle the critical point $(-1,\ j0)$, meaning that the system possess a robust stability property.
4) The final transfer function matrix $N(s) = C(sI - A)^{-1}B + D$ is such that $N(s) - N(-s)$ has no transmission zeros on the imaginary axis except possibly at s=0.

Likewise, a transfer function is Negative Imaginary (NI), if the following conditions are satisfied:

1) It has no poles in the open right half s-plane.
2) Its Nyquist figure locates below the real axis for the positive frequencies and touch, but does not cross this axis. This condition can be illustrated mathematically as follows: $N(s) \in \text{SNI}$ if $j[N(j\omega) - N(-j\omega)] \geq 0$ for all $\omega > 0$.

The main stability result in NI systems theory [30] indicates that a positive closed-loop interconnection between an SNI system and its NI or SNI controller $N(s)$ is internally stable if and only if the DC loop gain at the zero frequency is less than unity, meaning that $\lambda_{max}[M(0)N(0)]<1$, where λ_{max} is the maximum eigen-value.

It is also assumed that an SNI controller has a good disturbance rejection since its numerator (DC gain) can be set high so that the response of the system can approach its desired value. Meanwhile, the stability and transient response of the closed loop control system can be guaranteed by choosing an appropriate value of damping ratio as given in its denominator. For example, the first-order SNI controller is designed with a general form as follows

$$M(s) = \frac{\delta}{as + \omega^2} = \frac{\delta / \omega^2}{(a/\omega^2)s + 1} = \frac{\gamma}{1 + \tau s} \tag{8.16}$$

where a>0 is the damping constant relative to ω>0. δ points out the gain values of our controller. The transfer function in (8.16) gives the time constant $\tau = a/\omega^2$ and the DC gain $\gamma = \delta/\omega^2$.

Takagi-Sugeno Fuzzy Inference System

Given a set of fuzzy linguistic rules, in which the knowledge is manifested in the antecedent and consequent parts (IF-THEN fuzzy rules), the standard model of the Takagi-Sugeno (TS) Fuzzy Systems can be represented using the following equation:

$$u(kT) = \frac{\sum_{i=1}^{N} w_i(kT) z_i(kT)}{\sum_{i=1}^{N} w_i(kT)}, \forall k \geq 0 \tag{8.17}$$

where $w_i()$ indicates the output weight for a particular rule and $z_i()$ is the output level of a certain rule. Meanwhile, the normalised weights $w_i()$ with respect to the sum of the total weight of all rules give the firing strength (the degree-of-fulfilment) of a certain rule while T denotes the sampling time. The degree-of-fulfilment indicates the level to which the antecedent part of a fuzzy rule is satisfied [48] and [7].

To properly define fuzzy rules, we employ five membership functions, representing the linguistic variables of our FIS, namely, NB for Negative Big, NS for Negative Small, Z for Zero, PS for Positive Small, and PB for Positive Big. Meanwhile, we also define the output of our TS fuzzy system z_i, i=1,2,3 in three steps, namely, H for high, M for medium and L for low.

To simplify the design process, we first normalise the range of the input error of the SNI controller within $-1 \leq e(\cdot) \leq 1$ before assigning five Gaussian fuzzy membership functions as follows:

$$\mu_i(kT) = exp\frac{(-e(kT) - \sigma_i)^2}{2c_i^2}, i = 1, 2, .., N \tag{8.18}$$

where kT denotes the sampling time, σ_i is the width of the ith pulse, and c_i indicates the centre of each pulse.

We transform our design knowledge in the form of IF-THEN fuzzy rules (e.g., if the error is NB then the gain of the SNI controller $\gamma()$ is set to become H, if the error is S then the gain of the SNI controller $\gamma()$ is set to become L, etc.). This way, we can effectively schedule the gain and time constant of our SNI controller based on the dynamics of the error signal in order to minimise the overshoots. Accordingly, we set the fuzzy rules as a representation of human knowledge, accommodating the need of our SNI controller in real time.

Based on the predefined membership function (8.18) and the knowledge-based fuzzy rules employed, our FIS system (8.17) can now perform an automatic tuning in real-time to determine the appropriate parameters of our SNI controller in terms of its time constant and gain.

Adaptive Fuzzy-Strictly Negative Imaginary Formation Control Strategy Design

Combining the advantages of our hybrid SNI controller and Fuzzy Inference System, we introduce the adaptive SNI control system to improve the robustness of the formation control under atmospheric disturbances. The fuzzy logic controller can adjust the gains of the SNI controller online (denoted by $\gamma_{x,y,z}$ and $\tau_{x,y,z}$, respectively) in order to cope with the environmental variance. To do so, the distance errors with disturbance can be considered as the inputs of the pre-defined membership functions in the Fuzzy logic. According to the Fuzzy logic rules, the SNI parameters will then be produced. The closed-loop relative-position tracking control diagram for our adaptive formation control system along the two axes is given in Figure 8.11.

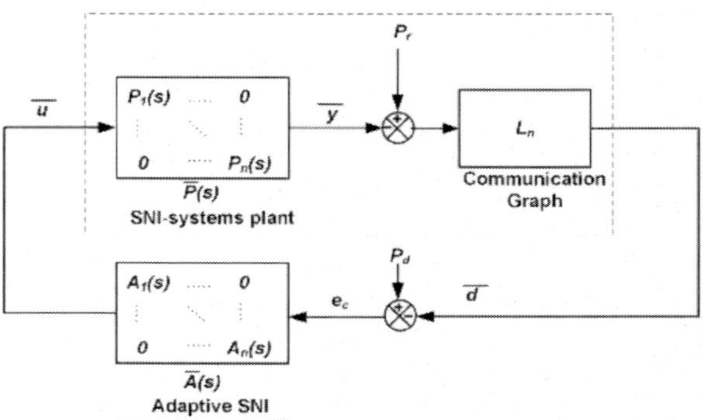

(a)

Figure 8.11 (Continued)

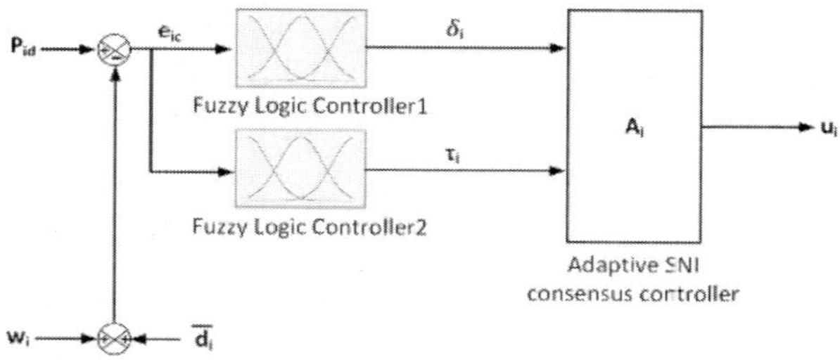

(b)

Figure 8.11. Diagram of the adaptive formation control system.
(a) Adaptive F-SNI formation control system based on the relative position consensus;
(b) Adaptive control diagram of each SNI consensus controller using Fuzzy Logic.

Simulation Tests without Added Wind Gusts

The dynamic models of the AR.DRone (3)-(5), illustrated in Section III, were implemented to test the results of the adaptive SNI leaderless formation control approach based on the relative position consensus control of multi-AR.Drones in the x and y (forward and lateral) directions.

Simulation Setup

The tuning of γ_{xy} and τ_{xy} in adaptive SNI controllers was based on the Takagi-Sugeno Fuzzy system, whose membership functions include 5 areas, namely, NB, NS, Z, PS, PB as shown in Figure 8.12. Their inputs are the normalized position errors $\in [-1\ 1]$ m on the x, y and z directions while their outputs are the $\gamma_{xy} \in [0.25\ 0.45]$ and $\tau_{xy} \in [0.5\ 1.5]$ for each axis.

Finally, an example was performed by considering the relative position consensus between UAVs to generate a square formation via our adaptive method.

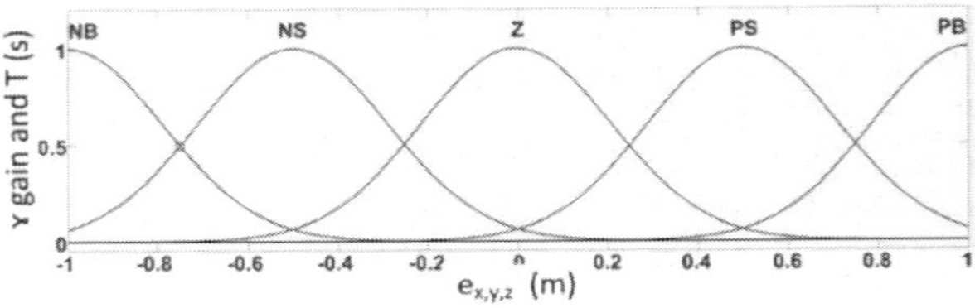

Figure 8.12. Membership function plots of γ_{xy} and τ_{xy}.

Simulation Outcomes

The desired formation is given in Figure 8.13. The transfer functions of four UAVs are used in this test. The communication topology can be seen in Figure 8.14 and hence

$$\mathcal{L}_n = \begin{bmatrix} 1 & -1 & 0 & 0 \\ -1 & 2 & -1 & 0 \\ 0 & -1 & 2 & -1 \\ 0 & 0 & -1 & 1 \end{bmatrix}$$

It can easily be seen that $\max_{(i=1)}^{4} \cdot \lambda_{\max}(N_{x_i}(0)) = 36.05 > 0$ for the x axis, where as, $\max_{(i=1)}^{4} \cdot \lambda_{\max}(N_{y_i}(0)) = 70.57 > 0$ for the y axis. Both the adaptive SNI controllers are selected as $\frac{-\gamma_{xy}}{1+\tau_{xy}}$, which has a negative DC gain value. Such that, $\lambda_{\max}[M_{xy_i}(0)N_{xy_i}(0)] < 0$ while $\frac{1}{\mathfrak{L}_n} = \frac{1}{3} \quad \forall i \in \{1,2,3,4\}$. In addition, $M_{xy_i}(\infty)N_{xy_i}(\infty) = 0 \quad \forall i \in \{1,2,3,4\}$. All satisfy the conditions described in ([32]) for an internal stability of the relative position consensus protocol for SNI UAV systems.

The output relative positions in the x and y (forward and lateral) directions of the system with the SNI adaptive consensus control method is shown in Figures 8.15 and 8.16 respectively. As can be seen, this controller is able to track its desired relative position reference input with satisfactory performance. It exhibits fast output responses, less overshoot, and also has a reasonable settling time at approximately 3s.

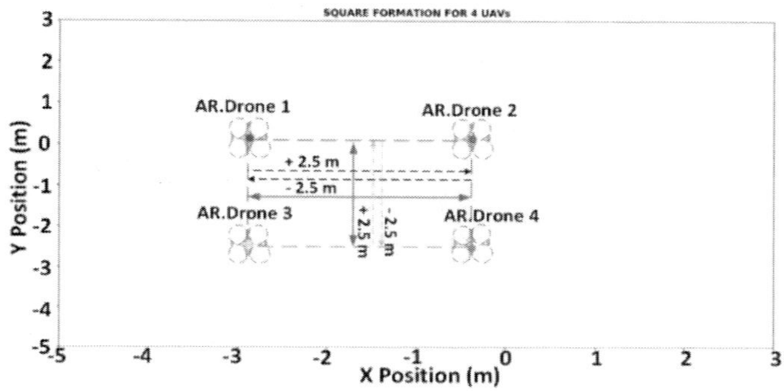

Figure 8.13. Expected square formation of 4 UAVs.

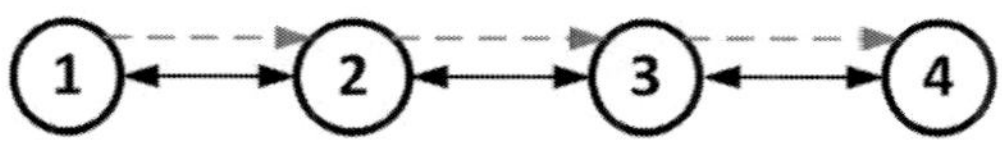

Figure 8.14. Communication graph for four UAVs.

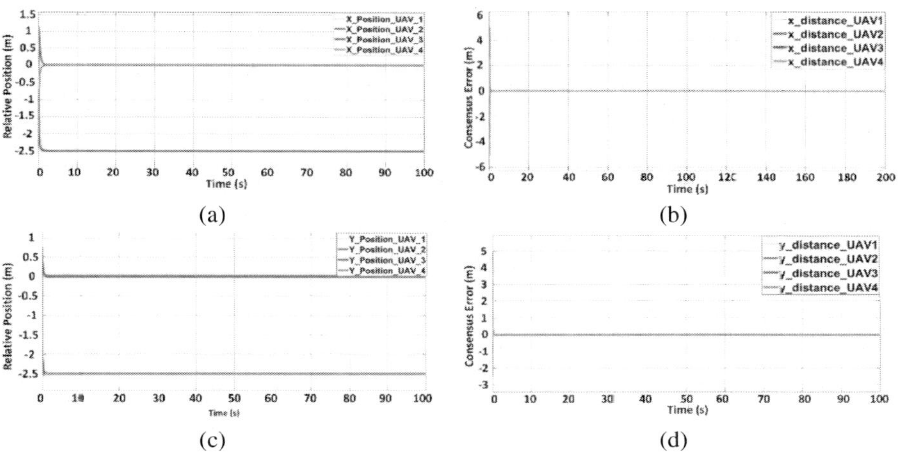

Figure 8.15. Relative position tracking results of 4 UAVs.
(a) actual relative position performance of 4 UAVs, (b) its consensus error on the *x* axis, (c) actual relative position performance of 4 UAVs and (d) its consensus error on the *y* axis.

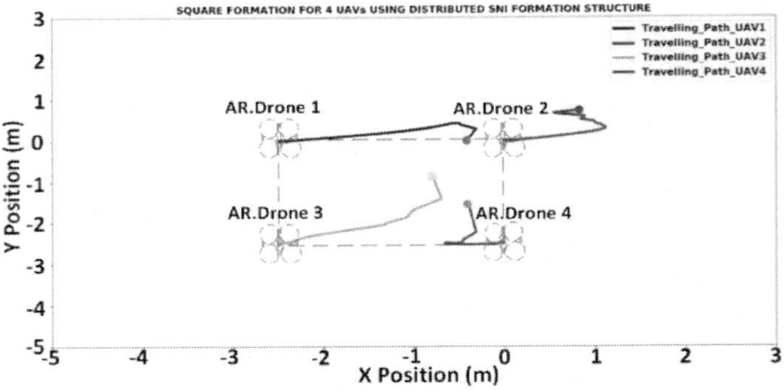

Figure 8.16. Traveled trajectories of 4 UAVs to achieve the given square formation.

Experimental Results with Wind Gust Disturbances

In addition to some noticeable advantages, such as low position errors, less overshoot and quick response as shown in simulation results, the F-SNI formation control approach without the leader UAV is also capable of adjusting consensus control gains to eliminate the uncertain nature of flight environments. Since the fixed-gain SNI is only a static controller whose γ and τ gains are set as constant values, its control actions cannot be improved by time. Hence, in this section, we validate the efficiency of our proposed controller in a multi-UAV flight experiment, being strongly affected by a simulated wind-gust source.

Experimental Configuration

We implemented our experiments in the indoor flight test laboratory. The flight zone was equipped with 17 VICON motion capture cameras with millimeter accuracy. The camera systems, which can simulate the function of relative position sensors, are used to measure the relative positions between two adjacent AR.Drone quadcopters. It is also assumed that each quadrotor can sense its orientation from a sensor such as an IMU so that they can share the same axes orientation.

The two comparative consensus controllers $A_i(s)$, including the PIDF and F-SNI, are tuned properly (e.g., minimizing the quadratic function (19)) to achieve the best performance for consensus control in the x and y directions.

$$E_{t_{x,y}} = \frac{1}{2}(e_{\delta_{k_{x,y}}}(kT))^2, \forall k > 0 \tag{8.19}$$

where k is the sampling time, $e_{\delta_{k_{x,y}}}$ is the relative position error, and T is the sampling period. The PID parameters after tuning process are given in (8.20)–(8.21).

$$P_{x_i PIDF}(s) = K_p + \frac{K_i}{s} + \frac{K_d Ns}{s+N} = 0.036 + \frac{0.0012}{s} - \frac{0.0084s}{s+0.055} \tag{8.20}$$

$$P_{y_i PIDF}(s) = P_{z_i PIDF}(s) = K_p + \frac{K_i}{s} + \frac{K_d Ns}{s+N} = 0.122 + \frac{0.0041}{s} - \frac{0.0084s}{s+0.065} \tag{8.21}$$

Additionally, a fan, located at the diagonal side of the positive x and y directions with a span of 1.8m, was used to mimic outside wind gusts. The fan produces a relative wind which alters in strength from the center to the outside of the fan exhaust and can deliver wind speeds of up to 5m/s. During our experiments, the fan is turned on at speed level No.2, which is equivalent to approximately 3m/s (see Figure 8.17).

All drones in our formation test are randomly placed on the ground. They then perform a vertical take-off and hover at an expected altitude of 1m for a duration of 10s (marked as circles on the UAV trajectories). After 10s, the F-SNI distributed formation controllers will be activated.

Figure 8.17. Formation flight conducted in our lab.

Experimental Results

Because of the limited test space and the sonar interference between AR.Drones, only two drones are employed in this test. The desired formation between 2 UAVs was presented in Figure 8.18, where the relative coordinate between two drones is (2.7, 0.5) m.

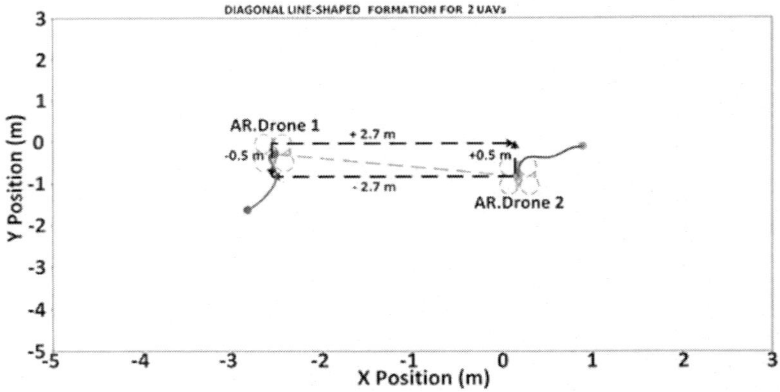

Figure 8.18. Expected diagonal-line formations of 2 UAVs in the presence of wind gusts.

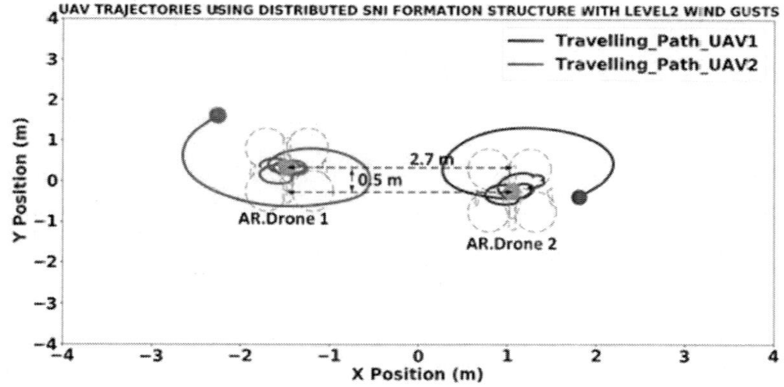

(a)

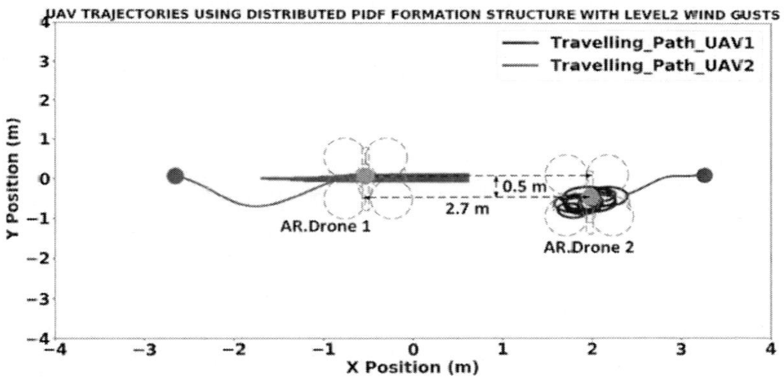

(b)

Figure 8.19. Comparative performances: (a) The formation tracking performance of the F-SNI controller and (b) The formation tracking performance of the PIDF controller.

As can be seen in Figure 8.19, the high-intensity wind disturbances degrade the performance of both controllers along two axes. However, our control method is capable of stabilizing the multi-UAV formation system without notable oscillations and reaching a strong relative position consensus with less drift, even in the presence of added wind gust disturbances (approximately 3m/s) in both directions (Figure 8.19a). The adaptive SNI's evaluation metrics yields much more desirable values, where the Root Mean Square Error (RMSE) value is only 0.15 m and a maximum variation (MV) is 0.3m. It is clear that the multi-UAV system with the F-SNI consensus controller was able to adequately compensate for wind gust disturbances. In contrast, given the same wind speeds, the PIDF controller was unable to produce a satisfactory performance by demonstrating a reasonably large RMSE value of 0.36m and a high MV of 1m (Figure 8.19b). Moreover, under the control of F-SNI controller, our quadrotors quickly create the desired leader-less formation in only 4s, which is twice as fast as that system with the PIDF. These comparative performances and metric values validate the capacity of our controller with respect to the fixed-gain or conventional PID control methods in improving the robustness of the closed-loop systems against the changing environmental conditions.

CONCLUSION

The simulation and real flight tests carried out have shown that using the F-SNI controller for graph theory-based multi-UAV relative position consensus control can achieve and maintain a global asymptotic convergence to the desired formation in the face of large uncertainties. Utilising the relative position information to each of the surrounding robots (which could, for example, be measured directly by the camera or laser scanner system mounted on each of the UAVs) instead of global state locations regularly utilized in the literature (which must be continuously shared with each other within the robot team through the Wi-Fi network), cooperative tasks for large-scale multi-UAV system can be efficiently and satisfactorily performed since the distributed control problems such as high computational demand and bandwidth congestion have never been encountered in our control approach.

Further, compared to the performance of the PIDF control strategies, the A-SNI formation control method outperforms the PIDF since its RMSE value is eight times lower than that of the PIDF controller. Overall, the performance of our adaptive SNI consensus controllers meet the goals of obtaining highly precise formation tracking for the multi-UAV system with a pre-defined communication graph while accommodating the unknown disturbances and keeping minimal steady-state error.

In future, this method will be combined with a trajectory planner to solve the problem of moving the multi-UAV formation to a target point. Further, a neural network technique, which has the capability of online learning, will be developed to automatically adjust the width of each Fuzzy membership function so that the F-SNI controller can adapt better with many various kinds of exogenous disturbances.

REFERENCES

[1] Sherstjuk, V., Zharikova, M., Sokol, I. (2018, April). Forest fire-fighting monitoring system based on uav team and remote sensing. In *Proceedings of 2018 IEEE 38th International Conference on Electronics and Nanotechnology* (pp. 663-668). IEEE.

[2] Ge, X., Han, Q. L., Zhang, X. M. (2017). Achieving cluster formation of multi-agent systems under aperiodic sampling and communication delays. *IEEE Transactions on Industrial Electronics*, 65(4), 3417-3426.

[3] Pandolfo, M. (2017). Modeling and control of a suspended load lifted by two aerial vehicles.

[4] Macwan, A., Vilela, J., Nejat, G., Benhabib, B. (2014). A multirobot path-planning strategy for autonomous wilderness search and rescue. *IEEE transactions on cybernetics*, 45(9), 1784-1797.

[5] Martin, M., & Kilberg, S. (2001). TECHSAT 21 and Revolutionizing Space Missions using Microsatellites, *American Institute of Aeronautics and Astronautics*. SSC-01-1-3.

[6] Santoso, F. (2014). Range-only distributed navigation protocol for uniform coverage in wireless sensor networks. *IET Wireless Sensor Systems*, 5(1), 20-30.

[7] Shladover, S. E. (2018). Connected and automated vehicle systems: Introduction and overview. *Journal of Intelligent Transportation Systems*, 22(3), 190-200.

[8] Keshmiri, S., Payandeh, S. (2009). A centralized framework to multi-robots formation control: Theory and application. In *Collaborative Agents-Research and Development* (pp. 85-98). Springer, Berlin, Heidelberg.

[9] Oleiwi, B. K., Al-Jarrah, R., Roth, H., Kazem, B. I. (2015). Integrated Motion Planning and Control for Multi Objectives Optimization and Multi Robots Navigation. *IFAC-PapersOnLine*, 48(10), 99-104.

[10] Peng, Z., Yang, S., Wen, G., Rahmani, A., Yu, Y. (2016). Adaptive distributed formation control for multiple nonholonomic wheeled mobile robots. *Neurocomputing*, 173, 1485-1494.

[11] Chen, Y., Sun, J. (2016). Distributed optimal control for multi-agent systems with obstacle avoidance. *Neurocomputing*, 173, 2014-2021.

[12] Tran, V. P., Garratt, M., Petersen, I. R. (2017, December). Formation control of multi-UAVs using negative-imaginary systems theory. In *Proceedings of 2017 11th Asian Control Conference* (pp. 2031-2036). IEEE.

[13] Tran, V. P., Garratt, M., & Petersen, I. R. (2020). Switching time-invariant formation control of a collaborative multi-agent system using negative imaginary systems theory. *Control Engineering Practice*, *95*, 104245.

[14] Dai, Y., Kim, Y., Wee, S., Lee, D., Lee, S. (2015). A switching formation strategy for obstacle avoidance of a multi-robot system based on robot priority model. *ISA transactions*, 56, 123-134.

[15] Kuo, C. W., Tsai, C. C., Lee, C. T. (2019). Intelligent leader-following consensus formation control using recurrent neural networks for small-size unmanned helicopters. *IEEE Transactions on Systems, Man, and Cybernetics: Systems* online first.

[16] Dong, X., Zhou, Y., Ren, Z., Zhong, Y. (2016). Time-varying formation tracking for second-order multi-agent systems subjected to switching topologies with application to quadrotor formation flying. *IEEE Transactions on Industrial Electronics*, 64(6), 5014-5024.

[17] Wang, R., Dong, X., Li, Q., Ren, Z. (2018, July). Fully Distributed Output-feedback Time-varying Formation Tracking Control for

Second-order Nonlinear Multi-agent Systems with Switching Directed Topologies. In *Proceedings of 37th Chinese Control Conference* (pp. 7088-7093). IEEE.

[18] Wei, H., Lv, Q., Duo, N., Wang, G., Liang, B. (2019). Consensus Algorithms Based Multi-Robot Formation Control under Noise and Time Delay Conditions. *Applied Sciences*, 9(5), 1004.

[19] Zhang, Y., Song, G., Qiao, G., Zhang, J., Peng, J. (2014, December). Consensus and obstacle avoidance for multi-robot systems with fixed and switching topologies. In *Proceedings of 2014 IEEE International Conference on Robotics and Biomimetics* (pp. 1213-1218). IEEE.

[20] Bhandari, S., Colgren, R. (2015). High-order dynamics models of a small UAV helicopter using analytical and parameter identification techniques. *Journal of the American Helicopter Society*, 60(2), 1-10.

[21] Santoso, F., Garratt, M. A., Anavatti, S. G., Petersen, I. (2018). Robust hybrid nonlinear control systems for the dynamics of a quadcopter drone. *IEEE Transactions on Systems, Man, and Cybernetics: Systems.*

[22] Wang, C., Tnunay, H., Zuo, Z., Lennox, B., Ding, Z. (2018). Fixed-time formation control of multirobot systems: Design and experiments. *IEEE Transactions on Industrial Electronics*, 66(8), 6292-6301.

[23] Galzi, D., Shtessel, Y. (2006, June). UAV formations control using high order sliding modes. In *Proceedings of 2006 American Control Conference* (pp. 6-pp). IEEE.

[24] Caudill, J. S., Harvey, M. B., Ashok, R. S. (2014, March). An overview of satellite formation control using higher order and adaptive sliding mode control techniques. In *Proceedings of IEEE SOUTHEASTCON*, 2014 (pp. 1-6). IEEE.

[25] Budiyanto, A., Cahyadi, A., Adji, T. B., Wahyunggoro, O. (2015, August). UAV obstacle avoidance using potential field under dynamic environment. In *Proceedings of 2015 International Conference on Control, Electronics, Renewable Energy and Communications* (pp. 187-192). IEEE.

[26] Wahyunggoro, O., Cahyadi, A. I. (2016). Path Planning Based on Fuzzy Decision Trees and Potential Field. *International Journal of Electrical & Computer Engineering*, 6(1).

[27] Schiano, F., Alonso-Mora, J., Rudin, K., Beardsley, P., Siegwart, R. Y., Sicilianok, B. (2014). Towards estimation and correction of wind effects on a quadrotor uav. In *Proceedings of International Micro Air Vehicle Conference and Competition* 2014 (pp. 134-141).

[28] Sikkel, L. N. C., de Croon, G. C. H. E., De Wagter, C., Chu, Q. P. (2016, October). A novel online model-based wind estimation approach for quadrotor micro air vehicles using low cost MEMS IMUs. In *Proceedings of 2016 IEEE/RSJ International Conference on Intelligent Robots and Systems* (pp. 2141-2146). IEEE.

[29] Fernández, R. A. S., Dominguez, S., Campoy, P. (2017, June). L_1 adaptive control for wind gust rejection in quad-rotor uav wind turbine inspection. In *Proceedings of 2017 International Conference on Unmanned Aircraft Systems* (pp. 1840-1849). IEEE.

[30] Petersen, I. R., Lanzon, A. (2010). Feedback control of negative-imaginary systems. *IEEE Control Systems Magazine*, 30(5), 54-72.

[31] Petersen, I. R. (2016). Negative imaginary systems theory and applications. *Annual Reviews in Control*, 42, 309-318.

[32] Wang, J., Lanzon, A., Petersen, I. R. (2015). Robust cooperative control of multiple heterogeneous negative-imaginary systems. *Automatica*, 61, 64-72.

[33] Wang, J., Lanzon, A., Petersen, I. R. (2015). Robust output feedback consensus for networked negative-imaginary systems. *IEEE Transactions on Automatic Control*, 60(9), 2547-2552.

[34] Perozzi, G., Efimov, D., Biannic, J. M., Planckaert, L., Coton, P. (2017, December). On sliding mode control design for UAV using realistic aerodynamic coefficients. In *Proceedings of IEEE 56th Annual Conference on Decision and Control* (pp. 5403-5408). IEEE.

[35] Anavatti, S. G., Santoso, F., Garratt, M. A. (2015, October). Progress in adaptive control systems: past, present, and future. In *Proceedings of 2015 International Conference on Advanced Mechatronics, Intelligent Manufacture, and Industrial Automation* (pp. 1-8). IEEE.

[36] Xiangxiang, F., Yang, Y., Tianping, Z., Weixing, Z. (2015, July). TS fuzzy modeling-based anti-disturbance tracking control for systems with nonlinear disturbances. In *Proceedings of 34th Chinese Control Conference* (pp. 628-633). IEEE.

[37] Wei, X., Chen, N. (2014). Composite hierarchical anti-disturbance control for nonlinear systems with DOBC and fuzzy control. *International Journal of Robust and Nonlinear Control*, 24(2), 362-373.

[38] Yi, Y., Fan, X. X., Zhang, T. P. (2016). Anti-disturbance tracking control for systems with nonlinear disturbances using T–S fuzzy modeling. *Neurocomputing*, 171, 1027-1037.

[39] Domingos, D., Camargo, G., Gomide, F. (2016). Autonomous fuzzy control and navigation of quadcopters. *IFAC-PapersOnLine*, 49(5), 73-78.

[40] Le Yu, Q. L. (2017). Unmanned Aerial Vehicle (UAV) Cooperative Mission Planning. *American Journal of Engineering and Technology Management*, 2(4), 36-44.

[41] Nguyen, S. D., Choi, S. B., Nguyen, Q. H. (2018). A new fuzzy-disturbance observer-enhanced sliding controller for vibration control of a train-car suspension with magneto-rheological dampers. *Mechanical Systems and Signal Processing*, 105, 447-466.

[42] Lanzon, A., Petersen, I. R. (2008). Stability robustness of a feedback interconnection of systems with negative imaginary frequency response. *IEEE Transactions on Automatic Control*, 53(4), 1042-1046.

[43] Bhikkaji, B., Moheimani, S. R., Petersen, I. R. (2011). A negative imaginary approach to modeling and control of a collocated structure. *IEEE/ASME Transactions on Mechatronics,* 17(4), 717-727.

[44] Mabrok, M. A., Kallapur, A. G., Petersen, I. R., Lanzon, A. (2013). Spectral conditions for negative imaginary systems with applications to nanopositioning. *IEEE/ASME Transactions on Mechatronics*, 19(3), 895-903.

[45] Das, S. K., Pota, H. R., Petersen, I. R. (2014). Resonant controller design for a piezoelectric tube scanner: A mixed negative-imaginary

and small-gain approach. *IEEE Transactions on Control Systems Technology*, 22(5), 1899-1906.

[46] Cai, C., Hagen, G. (2010). Stability analysis for a string of coupled stable subsystems with negative imaginary frequency response. *IEEE Transactions on Automatic Control*, 55(8), 1958-1963.

[47] Das, K., Ghose, D. (2015). Broadcast control mechanism for positional consensus in multiagent systems. *IEEE Transactions on Control Systems Technology*, 23(5), 1807-1826.

[48] Takagi, T., Sugeno, M. (1985). Fuzzy identification of systems and its applications to modeling and control. *IEEE transactions on systems, man, and cybernetics*, (1), 116-132.

ABOUT THE EDITOR

Dianwei Qian, PhD
Associate Professor
School of Control and Computer Engineering,
North China Electric Power University, Beijing, China
Email: dianwei.qian@gmail.com; dianwei.qian@ncepu.edu.cn

Dianwei Qian is currently an Associate Professor in the School of Control and Computer Engineering, The North China Electric Power University, Beijing, China. He received his BE degree from the Hohai University (HHU), Nanjing, China, in 2003, and PhD degree from Institute of Automation, Chinese Academy of Sciences (CAS), Beijing, China, in 2008. He also received his Master degree from the Northeastern University (NEU), Shenyang, China, in 2005. His research draws on the diverse control methods to aid in the analysis and design of complex dynamical systems.

INDEX

#

A

B

C

D

M

N

O

P

R

S

T

U

V

W